STUDY GUIDE

# Human Biology

## Concepts and Current Issues

### FOURTH EDITION

Judith Stewart

Community College of Southern Nevada

**PEARSON**
Benjamin Cummings

San Francisco  Boston  New York
Cape Town  Hong Kong  London  Madrid  Mexico City
Montreal  Munich  Paris  Singapore  Sydney  Tokyo  Toronto

*Senior Acquisitions Editor:* Deirdre McGill Espinoza
*Project Editors:* Marie Beaugureau and Nicole George-O'Brien
*Managing Editor:* Wendy Earl
*Production Editor:* Leslie Austin
*Proofreader:* Carole Quandt
*Compositor:* Cecelia G. Morales
*Cover Designer:* Yvo Riezebos Design
*Senior Manufacturing Buyer:* Stacey Weinberger
*Marketing Manager:* Gordon Lee

Copyright © 2008 Pearson Education, Inc., publishing as Pearson Benjamin Cummings, 1301 Sansome St., San Francisco, CA 94111. All rights reserved. Manufactured in the United States of America. This publication is protected by Copyright and permission should be obtained from the publisher prior to any prohibited reproduction, storage in a retrieval system, or transmission in any form or by any means, electronic, mechanical, photocopying, recording, or likewise. To obtain permission(s) to use material from this work, please submit a written request to Pearson Education, Inc., Permissions Department, 1900 E. Lake Ave., Glenview, IL 60025. For information regarding permissions, call (847) 486-2635.

Many of the designations used by manufacturers and sellers to distinguish their products are claimed as trademarks. Where those designations appear in this book, and the publisher was aware of a trademark claim, the designations have been printed in initial caps or all caps.

Pearson Benjamin Cummings™ is a trademark, in the U.S. and/or other countries, of Pearson Education, Inc. or its affiliates.

ISBN: 0-3215-0020-2

ISBN: 978-0-321-50020-5

1 2 3 4 5 6 7 8 9 10—TCS—11 10 09 08 07

www.aw-bc.com

**PEARSON**
**Benjamin Cummings**

# Contents

Preface v

| | | |
|---|---|---|
| **Chapter 1** | Human Biology, Science, and Society | 1 |
| **Chapter 2** | The Chemistry of Living Things | 14 |
| **Chapter 3** | Structure and Function of Cells | 30 |
| **Chapter 4** | From Cells to Organ Systems | 46 |
| **Chapter 5** | The Skeletal System | 63 |
| **Chapter 6** | The Muscular System | 77 |
| **Chapter 7** | Blood | 91 |
| **Chapter 8** | Heart and Blood Vessels | 105 |
| **Chapter 9** | The Immune System and Mechanisms of Defense | 122 |
| **Chapter 10** | The Respiratory System: Exchange of Gases | 136 |
| **Chapter 11** | The Nervous System: Integration and Control | 153 |
| **Chapter 12** | Sensory Mechanisms | 171 |
| **Chapter 13** | The Endocrine System | 185 |
| **Chapter 14** | The Digestive System and Nutrition | 201 |
| **Chapter 15** | The Urinary System | 217 |
| **Chapter 16** | Reproductive Systems | 231 |
| **Chapter 17** | Cell Reproduction and Differentiation | 252 |
| **Chapter 18** | Cancer: Uncontrolled Cell Division and Differentiation | 269 |
| **Chapter 19** | Genetics and Inheritance | 283 |
| **Chapter 20** | DNA Technology and Genetic Engineering | 298 |
| **Chapter 21** | Development and Aging | 311 |
| **Chapter 22** | Evolution and the Origins of Life | 327 |
| **Chapter 23** | Ecosystems and Populations | 340 |
| **Chapter 24** | Human Impacts, Biodiversity, and Environmental Issues | 354 |

# Preface

*"Argue for your limitations, and you shall have them."*

As you begin your study of science, one of the most important skills to acquire is the belief that you can succeed. Combined with commitment, adequate study time, practice, and utilization of available resources, you can master the concepts presented in your textbook. As you approach the Study Guide, be sure you have thoroughly read and studied the text material. Select a quiet location. Work through the exercises first without referring to the textbook or your notes. Use section references that accompany the exercises only to clarify a question you cannot answer, or have answered incorrectly. The *Study Guide* will be of little value if used as a worksheet, filling in answers while referring to the text. Use the *Study Guide* to practice applying what you have already learned through studying the textbook, and the Chapter Test to predict your readiness for classroom exams.

Each chapter of the *Study Guide* includes the following components:

**Chapter Summary and Objectives:** A list of 20 key concepts presented in the textbook.

**Exercises:** A combination of matching, fill-in-the-blank, labeling, completion, short answer, word choice, and crossword puzzles that test your understanding of the topic. Answers to most exercises are provided in the answer key located at the end of the *Study Guide*.

**Chapter Test:** A selection of 25 multiple-choice questions that assess your mastery of the topic and predict your performance on an exam. Take the Chapter Test only after you have successfully completed the exercises.

**Key Concept Review Questions:** The 20 key concepts listed at the beginning of each *Study Guide* chapter are rewritten as questions, assessing your comprehension of the main topics presented in the text. Answers to these questions are found by referring back to the Chapter Summary and Objectives.

Schedule time to study on a regular basis, preferably daily. You may choose to study an entire chapter first prior to working the exercises, or you may study the textbook in sections that correspond to the sections in the Study Guide. Either way, be sure to review the material several times to achieve a thorough understanding. Learning science can change your life, your perspective on many societal issues, and your ability to make informed choices. It all begins with you.

JUDY STEWART

# 1

# Human Biology, Science, and Society

## Chapter Summary and Key Concepts

*After reading and studying this chapter you should know the following:*

**Sections 1.1, 1.2**

1. Biology is the study of life.

2. Living things can be characterized by their molecular composition, energy requirements, cellular nature, homeostatic mechanisms, responsiveness, ability to grow and reproduce, and evolution.

3. A system of classification categorizes living things according to their characteristics.

4. As living organisms, humans are characterized by bipedalism, opposable thumbs, a large brain, and complex language skills.

5. Levels of biological organization exist in living things, beginning with atoms and becoming progressively more complex and inclusive.

**Sections 1.3, 1.4**

6. Scientific knowledge is information about the natural world and is gained by practicing the scientific method.

7. The scientific method is a series of steps that allow scientists to gather information about the natural world and test their ideas.

8. The scientific method begins with an observation and is followed by a hypothesis, a prediction, an experiment, data, and a conclusion.

9. A well-supported hypothesis becomes a theory.

10. Scientists publish their work in peer-reviewed journals to allow other scientists to verify their findings.

11. Scientific information is presented in many forms; it is important to be aware of different degrees of accuracy in each type of media.

### Sections 1.5, 1.6

12. Good scientists use the tools of critical thinking.
13. Informed individuals can become critical thinkers.
14. Science uses statistics to interpret data.
15. A graph is one effective way to present information.
16. Facts differ from anecdotes and conclusions because facts are verifiable.
17. Correlation is a pattern between two variables, and it does not necessarily indicate causality.
18. Technology is the application of science and has a great potential to improve our world.
19. Science is an explanation of observable events in a physical world; it cannot be used to explain or support feelings, opinions, or abstract ideas.
20. A knowledge of science helps individuals to make informed choices.

# Exercises

Complete the exercises for each section after you have read and studied the section. If you cannot answer some questions, or answer them incorrectly, return to the chapter and review this information. You may find it helpful to work on only one section at a time. When you have completed all sections, take the Chapter Test as an indicator of your mastery of this topic.

**1.1 The characteristics of life**

**1.2 How humans fit into the natural world**

## Matching

_____ 1. **science**  a. one of two domains comprised of bacteria

_____ 2. **biology**  b. the second smallest unit of classification

_____ 3. **metabolism**  c. a kingdom of life containing all single-celled prokaryotes

_____ 4. **cell**  d. a kingdom of multicellular eukaryotes that obtain energy by eating plants or animals

_____ 5. **homeostasis**  e. the study of life

_____ 6. *Monera*  f. organisms that can interbreed in nature and produce fertile offspring

_____ 7. *Eukarya*  g. a kingdom of multicellular eukaryotes that capture and utilize solar energy

_____ 8. *Animalia*  h. a process that allows plants to capture energy from sunlight

_____ 9. *Plantae*  i. a kingdom containing unicellular and simple multicellular eukaryotes

_____ 10. *Fungi*  j. a level of classification that encompasses kingdoms

_____ 11. **photosynthesis**  k. physical and chemical processes involving the energy transformations that maintain life

_____ 12. *Protista*  l. the maintenance of a constant internal environment in living things

_____ 13. **domain**  m. a kingdom of multicellular eukaryotes that obtain energy from decaying material

_____ 14. **Bacteria**  n. in the three-domain system of classification, this domain comprises all organisms with cells containing a nucleus

_____ 15. **species**  o. the smallest unit of life that exhibits all the characteristics of life

_____ 16. **genus**  p. a second domain containing bacteria in the three-domain classification system

_____ 17. **Archae**  q. the study of the natural world

## Fill-in-the-Blank

*Referenced sections are indicated in parentheses.*

18. The natural world includes all _____ and all _____. (1.1)

19. _____ cells have a nucleus, and _____ cells lack a nucleus. (1.1)

20. When observing the seven signs of life, both living and nonliving things can grow, but only living things can _____. (1.1)

21. Animals with a nerve cord and a backbone are called _____. (1.2)

22. Vertebrates with mammary glands are classified as _____. (1.2)

23. Humans, apes, and monkeys are all considered mammals and are further classified as _____. (1.2)

24. The ability to stand upright and walk on two legs is called _____. (1.2)

25. Humans have _____ thumbs, allowing them to grasp small objects with precision. (1.2)

26. Humans have a large _____ mass in relation to their body size. (1.2)

27. The capacity for complex _____ is a defining feature of being human. (1.2)

## Ordering

*Place the following levels of biological organization in the proper sequence, beginning with the simplest and ending with the most complex.*

a. tissue          d. organ system     g. population
b. organism        e. community        h. atom
c. cell            f. organ            i. ecosystem

28. _____  29. _____  30. _____
31. _____  32. _____  33. _____
34. _____  35. _____  36. _____

## Short Answer

37. List the 7 criteria accepted by scientists as characteristics of life.

38. When are the characteristics of life not applicable in describing living things?

39. How is a domain different from a kingdom?

40. List the four defining features of humans.

## Critical Thinking

41. An organelle is a small functional unit found within a cell. Why is a single organelle not considered to be alive?

42. In order to sustain life, human beings obtain energy and raw materials from both living and nonliving things. Provide an example of how you, as a human being, exhibit this characteristic each day.

43. Maintaining homeostasis and responding to the external environment are two characteristics of living things. How are these two characteristics similar and how are they different?

1.3 Science is both a body of knowledge and a process
1.4 Sources of scientific information vary in style and quality
1.5 Learning to be a critical thinker
1.6 The role of science in society

## Matching

\_\_\_\_ 1. **scientific knowledge**  a. a broad hypothesis that has been repeatedly tested and supported over time

\_\_\_\_ 2. **scientific method**  b. a false treatment given to a control group in an experiment

\_\_\_\_ 3. **hypothesis**  c. a factor that might vary during the course of an experiment

\_\_\_\_ 4. **experiment**  d. the application of science

\_\_\_\_ 5. **variable**  e. the group in a controlled experiment that is not exposed to the experimental treatment

\_\_\_\_ 6. **experimental group**  f. a questioning attitude required to practice science

\_\_\_\_ 7. **control group**  g. organized, reliable information about the natural world

\_\_\_\_ 8. **placebo**  h. a tentative statement about the natural world

\_\_\_\_ 9. **theory**  i. a carefully planned and executed test of a prediction

\_\_\_\_ 10. **skepticism**  j. the group in a controlled experiment that is exposed to the experimental treatment

\_\_\_\_ 11. **data**  k. the process used to acquire scientific knowledge

\_\_\_\_ 12. **technology**  l. numerical information obtained from an experiment

## Fill-in-the-Blank

*Referenced sections are indicated in parentheses.*

13. Science is a study of the natural world and includes both _____ and _____. (1.3)

14. When we use observations to make broad generalizations, we are using _____ reasoning. (1.3)

15. The process of gaining scientific knowledge is called the _____ _____. (1.3)

16. A hypothesis is a statement that can lead to _____ deductions. (1.3)

17. A working hypothesis is a testable _____. (1.3)

18. In _____ reasoning, a generalization is applied to specific cases. (1.3)

19. The variable of interest in an experiment is called the _____ variable. (1.3)

20. When the subjects in an experiment do not know who is in the control group and who is in the experimental group, this is a _____ study. (1.3)

21. When neither the subjects nor the administrators of an experiment know who is in the control group and who is in the experimental group, this is a _____ study. (1.3)

22. When a prediction is found to be false, the _____ must be modified. (1.3)

23. A hypothesis cannot be _____, it can only be _____ or _____. (1.3)

24. The scientific method can be described as a process of _____. (1.3)

25. Scientists publish their findings in _____ journals so that other scientists can examine their work. (1.3)

26. The range above and below a numerical average that scientists believe represents accurate data is called the _____ _____. (1.5)

27. A testimonial is an example of _____ evidence. (1.5)

28. Information that can be verified is a _____. (1.5)

29. A relationship between two variables is a _____, but it does not always mean that one variable _____ the other. (1.5)

30. Science is limited to the explanation of _____ events in the _____ world. (1.6)

## Completion

*Scientific information is reported through many types of media. For each type listed below, name one strength and one weakness of this resource for a general biology student. (There are no answers to this exercise in the answer key. Refer to Section 1.4 of your text to check your responses.)*

| Type of Media | Strengths | Weaknesses |
|---|---|---|
| Scientific journal | | |
| Popular science magazine | | |
| Newspaper | | |
| Web site ending in .com | | |
| Web site ending in .edu | | |

## Labeling

*In Figure 1.1, indicate which of the following terms apply to indicated areas on the graph.*

a. ordinate
b. abscissa
c. independent variable
d. dependent variable
e. standard error bar

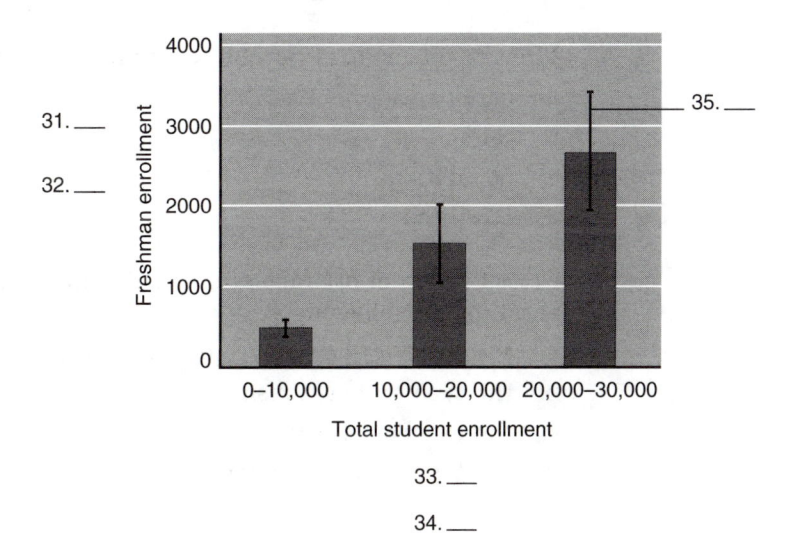

31. ___
32. ___
33. ___
34. ___
35. ___

**Figure 1.1**

# Chapter Test

## Multiple Choice

1. Biology is the study of:
   a. science.
   b. ecosystems.
   c. life.
   d. biochemistry.

2. Which of the following is the most accurate and inclusive list of what is contained in the natural world?
   a. all matter and all living things
   b. all matter and all nonliving things
   c. all energy and all living things
   d. all matter, all energy, all living things, and all nonliving things

3. Living things require energy. The processes involved in acquiring and transforming energy are collectively referred to as:
   a. chemistry.
   b. physical biochemistry.
   c. metabolism.
   d. catabolism.

4. The absence of _____ would indicate a substance is not a living organism.
   a. multicellular structure
   b. molecules of life
   c. a nucleus
   d. external movement

5. What is the smallest level of order at which evolution occurs?
   a. organism
   b. cell
   c. population
   d. ecosystem

6. Classification of organisms in the five kingdom systems depends on all of the following *except:*
   a. the presence or absence of a nucleus.
   b. the type of metabolism.
   c. the number of cells.
   d. the type of reproduction.

7. Which of the following is a domain that is described incorrectly?
    a. Prokarya—contains all prokaryotic organisms
    b. Eukarya—contains all eukaryotic organisms
    c. Archaea—contains primitive unicellular eukaryotes
    d. Bacteria—contains unicellular prokaryotes

8. The defining features of humans include all of the following *except:*
    a. opposable thumbs.
    b. bipedalism.
    c. sexual reproduction.
    d. large brain mass in relation to body size.

9. A group of tissues that carry out a specific function constitute a(n):
    a. cell.
    b. organism.
    c. organ.
    d. community.

10. Place the following steps of the scientific method in the correct sequence in which they would be used.
    I. formulate a hypothesis
    II. observe and generalize
    III. experiment
    IV. make a testable prediction
    V. modify hypothesis if necessary

    a. I, V, II, IV, III
    b. II, I, IV, III, V
    c. II, IV, III, I, V
    d. IV, I, II, III, V

11. In a scientific experiment, a placebo is given to the:
    a. experimental group.
    b. control group.
    c. both groups at different times.
    d. neither group.

12. You wish to test the effect of different amounts of a particular fertilizer on the growth of tomato plants. In your experiment, the controlled variable is:
    a. the type of fertilizer.
    b. plant growth.
    c. the amount of fertilizer.
    d. the type of tomato plant used.

13. The most accurate results will be obtained by a scientific study that:
    a. has a control group and an experimental group.
    b. is "double-blind."
    c. is "blind."
    d. identifies variables that may affect the outcome of the experiment.

14. Which of the following does not describe a prediction?
    a. an "if...then" statement
    b. a vague, general statement
    c. a testable statement
    d. all of the above describe a prediction

15. Confidence in a hypothesis is developed when:
    a. many scientists propose the hypothesis at the same time.
    b. the experiment designed to test the hypothesis has a control group and an experimental group.
    c. the results of an experiment support the hypothesis.
    d. many scientists have tried and failed to disprove the hypothesis.

16. The goal of science is to _____ what is _____.
    a. prove, truth
    b. prove, false
    c. disprove, truth
    d. disprove, false

17. Scientists most often publish their findings:
    a. in popular magazines.
    b. in peer-reviewed journals.
    c. in documentaries.
    d. on the Internet.

18. A fact is:
    a. a judgment.
    b. an interpretation.
    c. verifiable information.
    d. numerical data.

19. Statistics is the mathematics of organizing and interpreting:
    a. graphs.
    b. experiments.
    c. variables.
    d. data.

20. You begin eating broccoli every night of the week, and notice that you are more energetic. Which of the following would be the most accurate statement?
    a. Eating broccoli causes you to be more energetic.
    b. Eating broccoli is a good way to combat fatigue.
    c. Eating broccoli appears to be correlated with increased energy.
    d. Eating broccoli provides nutrients that are causing you to have more energy.

21. Developing the skills of a critical thinker will most help you to:
    a. understand information.
    b. question information.
    c. generate information.
    d. prove information.

22. Science cannot be used to explain:
    a. data.
    b. observations of the physical world.
    c. nonphysical events outside the natural world.
    d. science can be used to explain all of the above.

23. Anecdotal evidence often appears as:
    a. a double-blind study.
    b. a testimonial by a nonscientist.
    c. data.
    d. statistical evidence.

24. In a graph, the standard error is represented by:
    a. the ordinate.
    b. the abscissa.
    c. small lines that extend above and below the average number.
    d. bars or dots indicating the average of all data collected.

25. Understanding science and the scientific method can help individuals:
    a. make informed decisions.
    b. contribute to society.
    c. understand the impact of technology.
    d. evaluate the validity of nonphysical events that do not fall within the limits of the natural world.

## Key Concept Review Questions

*Each of the Key Concepts listed at the beginning of this chapter has been rewritten as a question below. After successively completing the study guide exercises and the Chapter Test, you should be able to answer each of these questions. Refer to the Key Concepts list at the beginning of this chapter to check your answers.*

1. What is biology?
2. What are the seven characteristics of living things?
3. How do classification systems categorize living things?
4. What are the characteristics of humans?
5. What are the levels of biological organization?
6. What is scientific knowledge, and how is it gained?
7. What is the scientific method?
8. What are the steps of the scientific method?
9. What happens to a well-supported hypothesis?
10. Where do scientists most often publish their work?
11. When reading or watching the presentation of scientific information, what should you be aware of?
12. What type of thinking is used by good scientists?
13. Is this type of thinking limited to scientists?
14. What is the value of statistics to science?
15. What is a graph?
16. How do facts differ from anecdotes and conclusions?
17. What is a correlation? How does it relate to causality?
18. What is technology? What is the value of technology?
19. What are the limitations of science?
20. How is a knowledge of science valuable to individuals?

## Answer Key

### Sections 1.1, 1.2

**1.**q; **2.**e; **3.**k; **4.**o; **5.**l; **6.**e; **7.**n; **8.**d; **9.**g; **10.**m; **11.**h; **12.**i; **13.**j; **14.**a; **15.**f; **16.**b; **17.**p; **18.** matter, energy; **19.** eukaryotic, prokaryotic; **20.** reproduce; **21.** vertebrates; **22.** mammals; **23.** primates; **24.** bipedalism; **25.** opposable; **26.** brain; **27.** language; **28.**h; **29.**c; **30.**a; **31.**f; **32.**d; **33.**b; **34.**g; **35.**e; **36.**i; **37.** Refer to section 1.1.; **38.** While individual organisms may not show every characteristic, populations of similar organisms do.; **39.** Domain encompasses kingdoms.; **40.** bipedalism, opposable thumbs, large brain, complex language ability; **41.** The smallest

unit of life, exhibiting the characteristics of life, is a cell; since an organelle is only a part of a cell it does not qualify as living.;
**42.** Every day human beings breathe and bring water into their bodies, obtaining raw materials such as oxygen and hydrogen atoms from these nonliving substances. Energy is obtained from the foods we eat, such as plant and animal material, which are living substances.;
**43.** Homeostasis and responding to the external environment are similar because both processes seek to provide conditions favorable to life. They are different because homeostasis manages your internal environment inside your body, while responding to the external environment helps you interact in a healthy way with the world outside your body.

## Sections 1.3, 1.4, 1.5, 1.6

**1.**g; **2.**k; **3.**h; **4.**i; **5.**c; **6.**j; **7.**e; **8.**b; **9.**a; **10.**f; **11.**l; **12.**d; **13.** knowledge, process; **14.** inductive; **15.** scientific method; **16.** testable; **17.** prediction; **18.** deductive; **19.** experimental; **20.** blind; **21.** double-blind; **22.** hypothesis; **23.** proven, supported, disproved; **24.** elimination; **25.** peer-reviewed; **26.** standard error; **27.** anecdotal; **28.** fact; **29.** correlation, causes; **30.** observable, natural; **31.**a; **32.**d; **33.**b; **34.**c; **35.**e

## Chapter Test

**1.**c; **2.**d; **3.**c; **4.**b; **5.**c; **6.**d; **7.**c; **8.**c; **9.**c; **10.**b; **11.**b; **12.**b; **13.**b; **14.**a; **15.**d; **16.**d; **17.**b; **18.**c; **19.**d; **20.**c; **21.**b; **22.**c; **23.**b; **24.**c; **25.**d

# 2

# The Chemistry of Living Things

## Chapter Summary and Key Concepts

*After reading and studying this chapter you should know the following:*

**Sections 2.1, 2.2**

1. Matter is composed of elements; elements are composed of atoms; atoms are composed of protons, neutrons, and electrons.

2. The atomic number represents the number of protons in an atom; the mass number represents the total number of protons and neutrons in an atom's nucleus.

3. Isotopes are atoms of an element that have more or less neutrons than typical atoms.

4. Energy is the ability to do work. Electrons have potential energy based on their position in an electron shell.

5. Chemical bonds form between atoms that need to fill their outermost occupied electron shell.

6. Covalent bonds form when atoms share electrons; ionic bonds form between oppositely charged ions; hydrogen bonds form between polar molecules.

**Sections 2.3, 2.4**

7. Water is an effective biological solvent because the water molecules are polar, and water is a liquid at body temperature.

8. Water helps to regulate body temperature because it can absorb and hold heat.

9. An acid is a molecule that donates $H^+$ to a solution. A base is a molecule that absorbs $H^+$ in a solution.

10. The pH scale is a measure of the hydrogen ion concentration of a solution.

11. A buffer helps to minimize changes in pH.

### Sections 2.5, 2.6, 2.7, 2.8, 2.9, 2.10

12. Organic molecules are molecules that contain carbon covalently bonded to other elements.
13. Dehydration synthesis creates macromolecules by removing water. Hydrolysis breaks down macromolecules by adding water.
14. The four classes of organic molecules synthesized by living organisms are carbohydrates, lipids, proteins, and nucleic acids.
15. Carbohydrates in the form of simple sugars provide immediate energy to the cells, while polysaccharides provide stored energy.
16. Lipids include triglycerides that store energy, phospholipids that function in the cell membrane, and steroids that are classified as lipids because they are not soluble in water.
17. Proteins are composed of amino acids and are described at four structural levels.
18. Enzymes are proteins that function as catalysts to speed up chemical reactions.
19. Nucleic acids are composed of nucleotides.
20. DNA and RNA are nucleic acids that function in storing and expressing genetic information. ATP is a nucleic acid that functions as an energy molecule in the cell.

# Exercises

*Complete the exercises for each section after you have read and studied the section. If you cannot answer some questions, or answer them incorrectly, return to the chapter and review this information. You may find it helpful to work on only one section at a time. When you have completed all sections, take the Chapter Test as an indicator of your mastery of this topic.*

**2.1 All matter consists of elements**

**2.2 Atoms combine to form molecules**

## Matching

_____ 1. **matter**           a. atoms of an element with the same atomic number but a different atomic mass

_____ 2. **element**          b. stored energy that is not currently doing work

_____ 3. **atom**             c. a weak attractive force between oppositely charged regions of polar molecules containing hydrogen

_____ 4. **protons**          d. anything that has mass and occupies space

_____ 5. **neutrons**         e. enzymes and nutrients that protect the body from the action of free radicals

_____ 6. **electron**    f. a pure form of matter that cannot be broken down to a simpler form

_____ 7. **isotope**    g. energy that is doing work, the energy of motion

_____ 8. **molecule**    h. a bond formed by the sharing of electrons

_____ 9. **energy**    i. the smallest unit of any element that still retains the physical and chemical properties of that element

_____ 10. **potential energy**    j. attractive forces that bind atoms together

_____ 11. **kinetic energy**    k. neutral particles found in the nucleus of an atom

_____ 12. **chemical bonds**    l. the area within a shell where an electron is likely to be found

_____ 13. **covalent bond**    m. positively charged particles found in the nucleus of an atom

_____ 14. **ion**    n. the capacity to do work

_____ 15. **ionic bond**    o. negatively charged particles found orbiting around the nucleus of an atom

_____ 16. **hydrogen bond**    p. a stable association between two or more atoms

_____ 17. **orbital**    q. a bond formed by the attraction of two oppositely charged ions

_____ 18. **antioxidant**    r. an electrically charged atom

## Fill-in-the-Blank

*Referenced sections are indicated in parentheses.*

19. The nucleus of an atom contains _____ and _____. (2.1)

20. The second shell of an atom can hold up to _____ electrons. (2.1)

21. The _____ _____ represents the total number of neutrons and protons in the nucleus. (2.1)

22. All atoms of the same element have the same number of _____. (2.1)

23. _____ are unstable isotopes that give off energy in the form of radiation. (2.1)

24. The chemical formula $H_2S$ indicates that 1 molecule of this substance will have _____ atom(s) of hydrogen and _____ atom(s) of sulfur. (2.1)

25. When work is being done, _____ energy is converted to _____ energy. (2.2)

26. Electrons occupy an area within a shell called a(n) _____. (2.2)

27. Ions in solution may be called _____. (2.2)

28. _____ molecules are electrically neutral overall, but they have partially charged ends. (2.2)

29. The six most common elements in the body are _____, _____, _____, _____, _____, and _____. (2.2)

## Completion

*For each substance listed in the columns below, complete the table with required information (refer to Table 2.2 in the text).*

|  | # protons | # electrons | # neutrons | # electrons in the 2nd shell | # electrons in the 3rd shell | Is this an ion? |
|---|---|---|---|---|---|---|
| 30. Na | a. | b. | c. | d. | e. | f. |
| 31. Na$^+$ | a. | b. | c. | d. | e. | f. |
| 32. H | a. | b. | c. | d. | e. | f. |
| 33. H$^+$ | a. | b. | c. | d. | e. | f. |
| 34. C$_{12}$ | a. | b. | c. | d. | e. | f. |
| 35. N | a. | b. | c. | d. | e. | f. |
| 36. C$_{14}$ | a. | b. | c. | d. | e. | f. |

## Critical Thinking

37. Do double bonds ever occur between atoms joined by ionic bonds? Explain your answer.

38. a. What types of chemical bonds will exist in a glass of pure water?

    b. What types of chemical bonds will exist in a glass of salt water if some of the salt remains undissolved?

39. Is $H_2$ a polar molecule? Explain your answer.

2.3    Life depends on water

2.4    The importance of hydrogen ions

## Matching

_____ 1. **solvent**           a. a molecule that can give up a hydrogen ion

_____ 2. **solute**            b. a molecule that can accept a hydrogen ion

_____ 3. **acid**              c. any substance that minimizes changes in the pH of a solution when an acid or base is added

_____ 4. **base**              d. molecules that are polar and are attracted to water

_____ 5. **pH scale**          e. a dissolved substance

_____ 6. **buffer**            f. molecules that are not polar and are not attracted to water

_____ 7. **hydrophilic**       g. a liquid in which other substances dissolve

_____ 8. **hydrophobic**       h. a measure of $H^+$ concentration in a solution

## Completion

*Indicate which words or phrases below correspond to Figure 2.2.*

a. acidic solution
b. alkaline solution
c. neutral solution
d. higher $H^+$ concentration than pure water
e. same $H^+$ concentration as pure water
f. lower $H^+$ concentration than pure water

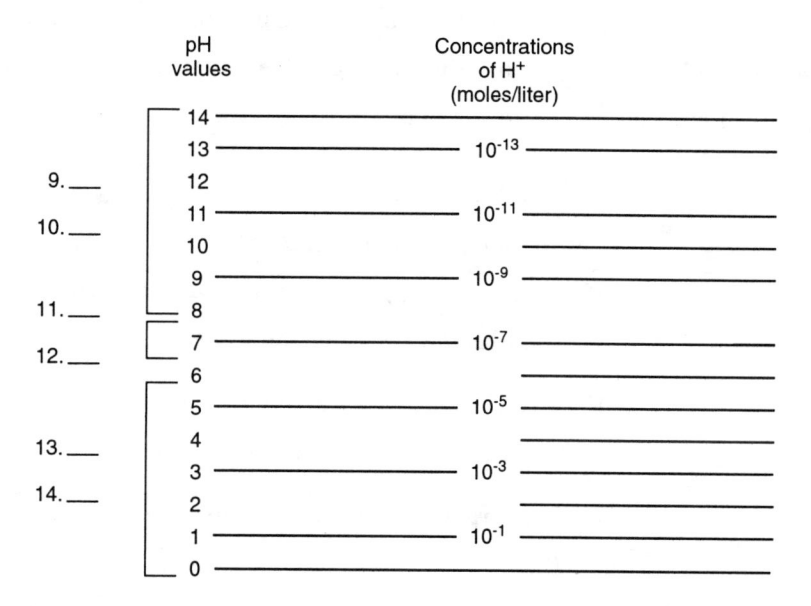

9. ___
10. ___
11. ___
12. ___
13. ___
14. ___

**Figure 2.2**

2.5  The organic molecules of living organisms
2.6  Carbohydrates: Used for energy and structural support
2.7  Lipids: Insoluble in water

## Matching

___ 1. **organic molecule**   a. a long straight or branched chain containing thousands of monosaccharides

___ 2. **dehydration synthesis**   b. a polysaccharide used by plants for structural support

___ 3. **hydrolysis**   c. the class of biochemical compounds that contains triglycerides, phospholipids, and steroids

___ 4. **monosaccharide**   d. a group of molecules classified as lipids because they are insoluble in water

___ 5. **oligosaccharide**   e. a molecule that contains carbon covalently bonded to other elements

___ 6. **polysaccharide**   f. a steroid that forms part of the cell membrane

____ 7. **glycogen**    g. a reaction in which small molecules are covalently bonded by the removal of a water molecule

____ 8. **starch**    h. a lipid composed of one molecule of glycerol and three fatty acids

____ 9. **cellulose**    i. a reaction in which macromolecules are broken apart by the addition of a water molecule

____ 10. **lipid**    j. a lipid that forms an important part of the cell membrane

____ 11. **triglyceride**    k. a short string of monosaccharides linked together

____ 12. **fatty acid**    l. the storage polysaccharide found in plants

____ 13. **phospholipid**    m. the simplest carbohydrate, consisting of a single sugar unit

____ 14. **steroid**    n. the storage polysaccharide found in animals

____ 15. **cholesterol**    o. a chain of hydrocarbons with a terminal carboxyl group

## Fill-in-the-Blank

*Referenced sections are indicated in parentheses.*

16. _____ are large organic molecules consisting of thousands of smaller molecules. (2.5)

17. The synthesis of macromolecules from smaller molecules _____ energy. (2.5)

18. The breakdown of macromolecules _____ energy. (2.5)

19. The four classes of organic molecules made by living organisms are _____, _____, _____, and _____. (2.5)

20. _____ is a six-carbon monosaccharide that serves as an important source of energy for cells. (2.6)

## Labeling

*Use the words below to identify each molecule or molecular region indicated in Figure 2.3.*

a. glycerol
b. steroid
c. monosaccharide
d. fatty acids
e. unsaturated fatty acid
f. triglyceride
g. disaccharide
h. polar head
i. saturated fatty acid
j. phospholipid
k. fatty acid tails

Chapter 2 *The Chemistry of Living Things* 21

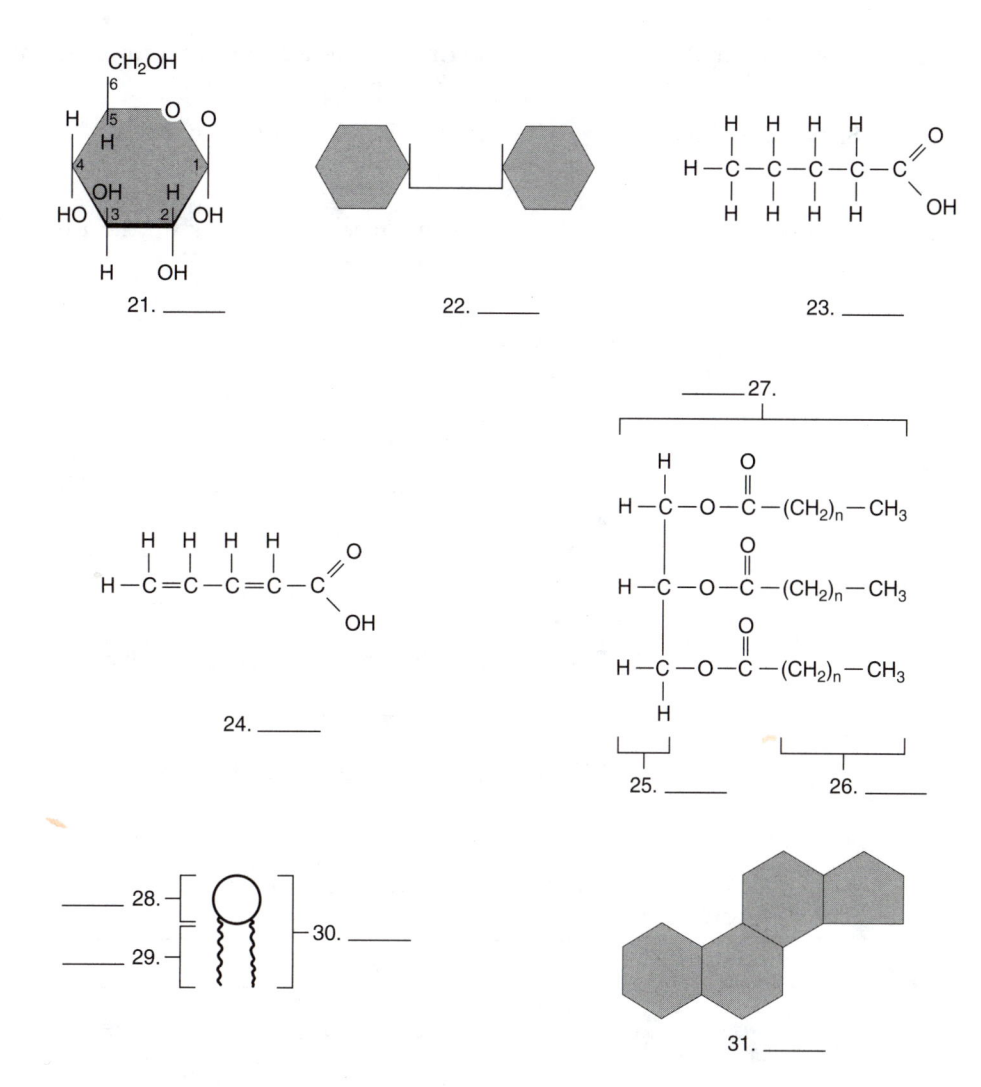

21. _____   22. _____   23. _____

27. _____

24. _____

25. _____   26. _____

28. _____
29. _____   30. _____

31. _____

**Figure 2.3**

2.8   **Proteins: Complex structures constructed of amino acids**

2.9   **Nucleic acids store genetic information**

2.10  **ATP carries energy**

# Matching

_____ 1. **amino acids**        a. a molecule produced by the breakdown of ATP

_____ 2. **polypeptide**        b. single units joined by covalent bonds in DNA and/or RNA

_____ 3. **protein**            c. a protein that speeds up a chemical reaction

_____ 4. **denaturation**       d. a nucleotide that functions as the universal energy source for cells

_____ 5. **enzyme**             e. single units joined by covalent bonds to form a protein

_____ 6. **catalyst**           f. a chain of more than 100 amino acids

## 22 Study Guide for *Human Biology*

____ 7. **DNA, RNA**

____ 8. **nucleotides**

____ 9. **ATP**

____ 10. **ADP**

g. permanent damage to protein structure resulting in loss of function

h. a string of 3–100 amino acids

i. nucleic acids

j. a substance that speeds up the rate of a chemical reaction

### Fill-in-the-Blank

*Referenced sections are indicated in parentheses.*

11. An enzyme can facilitate a reaction in which two reactants are joined to form a _____. (2.8)

12. A nucleotide consists of a _____, a _____, and a _____. (2.9)

13. DNA contains the instructions for producing _____. (2.9)

14. RNA contains the instructions for producing _____. (2.9)

15. The four bases found in DNA nucleotides are _____, _____, _____, and _____. (2.9)

16. RNA is a _____-stranded molecule containing a _____ sugar. (2.9)

17. The breakdown of ATP produces _____, _____, and _____. (2.10)

### Paragraph Completion

*Use the following terms to complete the paragraph, and then repeat this exercise with the terms covered.*

| | | | |
|---|---|---|---|
| beta-pleated sheet | denaturation | covalent | tertiary |
| amino | primary | temperature | R |
| carboxylic acid | secondary | amino acids | disulfide |
| structure | quaternary | pH | polar |

Proteins are macromolecules consisting of long chains of (18) _____ joined by (19) _____ bonds. Each amino acid has a central C-H group, bonded to an (20) _____ group, a (21) _____ group, and an additional (22) _____ group representing one of 20 different chemical possibilities. Protein function depends on the protein's (23) _____. The (24) _____ structure of a protein is the sequence of amino acids. Hydrogen bonding between amino acids at regular intervals produces the (25) _____ structure. Examples of this structure are the alpha-helix and the (26) _____. Twisting and folding of a protein into a three-dimensional shape produces the (27) _____ structure. This is the result of hydrogen bonds and (28) _____ bonds. Some proteins also have a (29) _____ structure, produced when two or more polypeptides are associated with each other. Because protein structure is maintained by weak hydrogen bonds, proteins can temporarily change their shape in response to (30) _____ molecules. A change in protein shape that leads to loss of function is called (31) _____ and may be caused by high (32) _____ or changes in (33) _____.

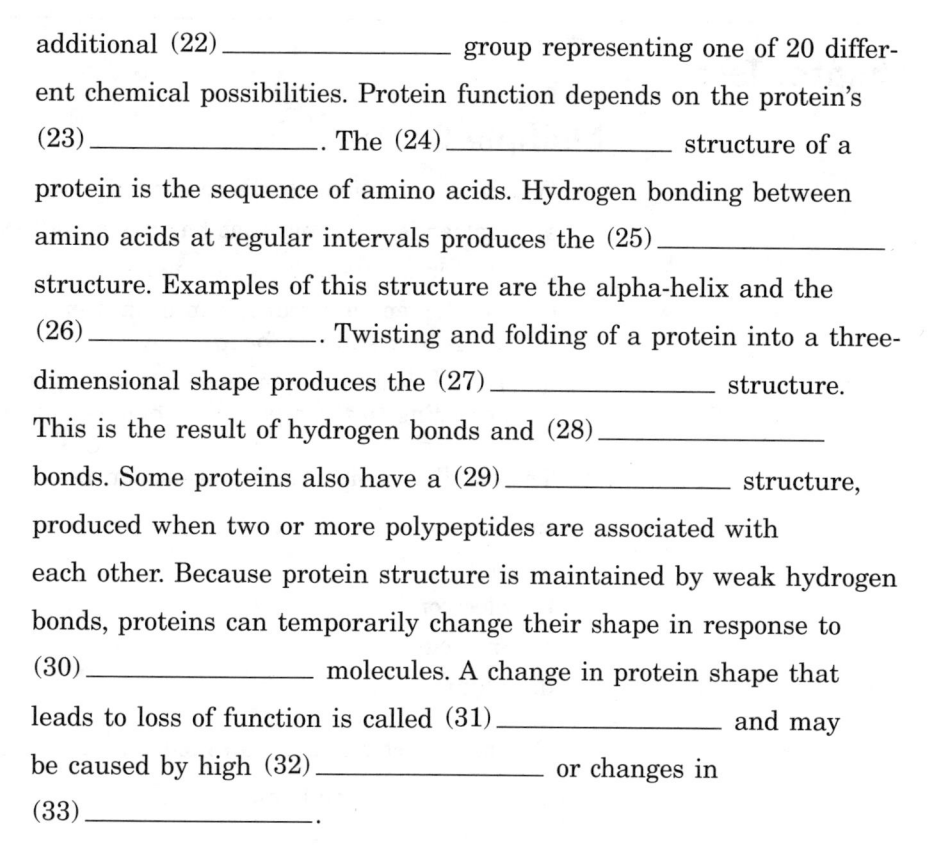

## Labeling

*Use the terms below to identify each molecule or molecular region indicated in Figure 2.4.*

a. triphosphate  c. R group  e. ribose  g. ATP
b. amino group  d. adenine  f. carboxylic acid group  h. amino acid

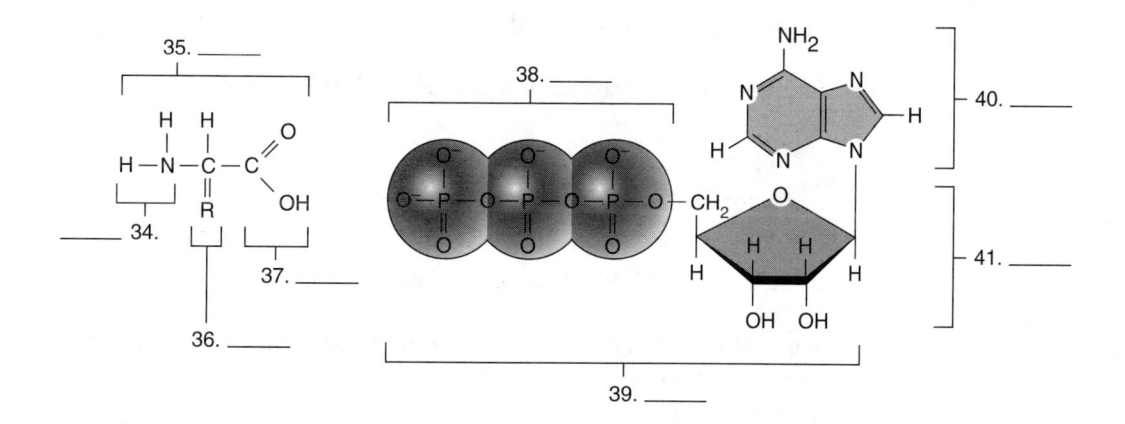

**Figure 2.4**

# Chapter Test

## Multiple Choice

1. Which of the following statements about matter is false?
   a. Matter is anything that has mass and occupies space.
   b. Matter is composed of elements.
   c. An element is a pure form of matter that cannot be broken down to a simpler form.
   d. The periodic table of elements arranges all known elements according to their physical chemical properties.

2. The smallest unit of matter that can take part in a chemical reaction is a(n):
   a. cell.
   b. electron.
   c. molecule.
   d. atom.

3. The first electron shell can hold _____ electrons, and the second shell can hold _____ electrons.
   a. 2, 2
   b. 2, 8
   c. 8, 8
   d. 2, 10

4. Atoms are electrically neutral when they have equal numbers of:
   a. protons and neutrons.
   b. neutrons and electrons.
   c. protons and electrons.
   d. protons, neutrons, and electrons.

5. Sodium has an atomic number of 11. This indicates that sodium has:
   a. 11 neutrons.
   b. a total of 11 particles in the nucleus.
   c. 11 protons.
   d. a total of 11 particles in the entire atom.

6. Most atoms of hydrogen have no neutrons. An atom of hydrogen containing one neutron would:
   a. be an ion.
   b. be an isotope.
   c. be a molecule.
   d. also have an additional proton.

7. One atom of C-12 has an atomic number of 6 and a mass number of 12. Which of the following is a true statement about C-14?
   a. C-14 has fewer protons than C-12
   b. C-14 has more neutrons than C-12
   c. C-12 has greater mass than C-14
   d. C-12 is neutral, while C-14 carries a charge

8. Atoms are most stable when:
   a. they have the same number of protons as neutrons.
   b. they have the same number of protons as electrons.
   c. their outermost occupied electron shell is partially filled.
   d. their outermost occupied electron shell is full.

9. Covalent bonds involve _____ electrons, and ionic bonds involve _____ electrons.
   a. transferring, sharing
   b. transferring, transferring
   c. sharing, transferring
   d. sharing, sharing

10. Radioisotopes are *not* used to:
    a. trace the movement of molecules in the body.
    b. trigger the movement of electrons between orbitals.
    c. destroy some types of cancer cells.
    d. determine the age of some rocks and fossils.

11. When an atom loses an electron, it becomes:
    a. positively charged.
    b. negatively charged.
    c. neutral.
    d. an isotope.

12. Which of the following is a FALSE statement about energy?
    a. energy is required to join atoms and form molecules, but not to break atoms within a molecule apart
    b. chemical bonds contain potential energy
    c. work is performed using kinetic energy
    d. electrons in the third shell of an atom have more potential energy than electrons in the second shell

13. The covalent bonds between oxygen and hydrogen in a water molecule:
    a. cannot be broken.
    b. can be broken, producing 2 hydrogen atoms and 1 oxygen atom.
    c. can be broken, producing 2 $H^+$ and 1 $O^-$.
    d. can be broken, producing 1 $H^+$ and 1 $OH^-$.

14. Which of the following occurs in a hydrogen bond?
    a. a positive charge hydrogen atom in one water molecule is attracted to a negatively charged oxygen atom in another water molecule
    b. the electrons in the hydrogen bond spend more time orbiting the oxygen atom than the hydrogen atom
    c. the electrons in the bond spend more time orbiting the hydrogen atom than the oxygen atom
    d. a and b
    e. a and c

15. An acid is a molecule that _____ $H^+$, while a base is a molecule that _____ $H^+$.
    a. absorbs, releases
    b. releases, absorbs
    c. reacts with, does not react with
    d. decreases, increases

16. Blood pH is maintained by the carbonic acid/bicarbonate buffer system: $HCO_3^- + H^+ \leftrightarrow H_2CO_3$. If the blood becomes too acidic, which of the following will occur?
    a. Carbonic acid will release $H^+$.
    b. Carbonic acid will absorb $H^+$.
    c. Bicarbonate will release $H^+$.
    d. Bicarbonate will absorb $H^+$.

17. The synthesis of macromolecules from smaller molecules:
    a. releases energy.
    b. requires energy.
    c. is accomplished by hydrolysis.
    d. requires the addition of water.

18. Monosaccharides:
    a. contain C, H, and O in a 1–2–1 ratio.
    b. contain C and H in a 1–1 ratio.
    c. contain equal amounts of C and H, with a small amount of O.
    d. contain equal amounts of C and O, with a small amount of H.

19. Which of the following is a carbohydrate with structural importance in plants?
    a. glycogen
    b. glucose
    c. starch
    d. cellulose

20. Triglycerides with a full complement of two H atoms for each C atom in their fatty-acid tails are:
    a. phospholipids.
    b. steroids.
    c. saturated.
    d. unsaturated.

21. Proteins function as catalysts when they:
    a. act as enzymes.
    b. speed up the rate of a chemical reaction.
    c. store genetic information.
    d. a and b

22. The three-dimensional protein shape maintained by hydrogen and disulfide bonds is the:
    a. primary structure.
    b. secondary structure.
    c. tertiary structure.
    d. quaternary structure.

23. What molecule is composed of a sugar, a phosphate, and a base?
    a. an amino acid
    b. a nucleotide
    c. a quaternary protein
    d. a saturated fat

24. Which of the following is *not* different in RNA and DNA?
    a. the phosphate group
    b. the sugar
    c. the presence of uracil
    d. the number of strands

25. The breakdown of ATP produces:
    a. ADP.
    b. free phosphate.
    c. energy.
    d. all of the above.

# Key Concept Review Questions

*Each of the Key Concepts listed at the beginning of this chapter has been rewritten as a question below. After successfully completing the study guide exercises and the Chapter Test, you should be able to answer each of these questions. Refer to the Key Concepts list at the beginning of this chapter to check your answers.*

1. What is matter composed of? What are elements composed of? What are atoms composed of?
2. What is indicated by the atomic number? What is indicated by the mass number?
3. What is an isotope?
4. What is energy? What type of energy is possessed by electrons based on their position?
5. What atoms tend to form chemical bonds?
6. Name three main types of bonds and describe how each is formed.
7. What characteristics of water make it a good biological solvent?
8. How does water help to regulate body temperature?
9. What is an acid? What is a base?
10. What is the pH scale?
11. What is the function of a buffer?
12. How can organic molecules be described?
13. How does dehydration synthesis differ from hydrolysis?
14. What are the four classes of organic molecules synthesized by living organisms?
15. Simple sugars and polysaccharides are both involved in energy; how are their roles different?
16. What are the three types of lipids and their functions?
17. What are proteins composed of? How are proteins characterized?
18. What is an enzyme?
19. What are nucleic acids composed of?
20. What are the functions of DNA, RNA, and ATP?

# Answer Key

## Sections 2.1, 2.2

**1.**d; **2.**f; **3.**i; **4.**m; **5.**k; **6.**o; **7.**a; **8.**p; **9.**n; **10.**b; **11.**g; **12.**j; **13.**h; **14.**r; **15.**q; **16.**c; **17.**l; **18.**e; **19.** neutrons, protons; **20.** two; **21.** mass number; **22.** protons; **23.** radioisotopes; **24.** two, one; **25.** potential, kinetic; **26.** orbital; **27.** electrolytes; **28.** Polar; **29.** oxygen, carbon, hydrogen, nitrogen, calcium, phosphorus; **30.** a.11, b.0, c.0, d.0, e.0, f.yes; **31.** a.11, b.10, c.12, d.8, e.1, f.yes; **32.** a.1, b.1, c.0, d.0, e.0, f.no; **33.** a.1, b.0, c.0, d.0, e.0, f.yes; **34.** a.6, b.6, c.6, d.4, e.0, f.no; **35.** a.7, b.7, c.7, d.5, e.0, f.no; **36.** a.6, b.6, c.8, d.4, e.0, f.no; **37.** No. Double bonds occur when atoms share two pairs of electrons, and there is no electron sharing in an ionic bond.; **38.** a. Covalent bonds (within each water molecule) and hydrogen bonds (between water molecules). b. Covalent bonds, hydrogen bonds, and ionic bonds between $Na^+$ and $Cl^-$.; **39.** No. Bonds formed between two atoms of the same element are covalent bonds. Both hydrogen atoms are identical, and so they have the same attraction for the shared electrons in the covalent bond. This means the electrons will not spend more time orbiting one nucleus than the other, and the molecule will not be polar.

## Sections 2.3, 2.4

**1.**g; **2.**e; **3.**a; **4.**b; **5.**h; **6.**d; **7.**c; **8.**f; **9.**b; **10.**f; **11.**c; **12.**e; **13.**a; **14.**d

## Sections 2.5, 2.6, 2.7

**1.**e; **2.**g; **3.**i; **4.**m; **5.**k; **6.**a; **7.**n; **8.**l; **9.**b; **10.**c; **11.**h; **12.**o; **13.**j; **14.**d; **15.**f; **16.** Macromolecules; **17.** requires; **18.** releases; **19.** carbohydrates, lipids, proteins, nucleic acids; **20.** Glucose; **21.**c; **22.**g; **23.**i; **24.**e; **25.**a; **26.**d; **27.**f; **28.**h; **29.**k; **30.**j; **31.**b

## Sections 2.8, 2.9

**1.**e; **2.**h; **3.**f; **4.**g; **5.**c; **6.**j; **7.**i; **8.**b; **9.**d; **10.**a; **11.** product; **12.** sugar, phosphate, base; **13.** RNA; **14.** protein; **15.** adenine, thymine, cytosine, guanine; **16.** single, ribose; **17.** ADP, P, energy; **18.** amino acids; **19.** covalent; **20.** amino; **21.** carboxylic acid; **22.** R; **23.** structure; **24.** primary; **25.** secondary; **26.** beta-pleated sheet; **27.** tertiary; **28.** disulfide; **29.** quaternary; **30.** polar; **31.** denaturation; **32.** temperature; **33.** pH; **34.**b; **35.**h; **36.**c; **37.**f; **38.**a; **39.**g; **40.**d; **41.**e

## Chapter Test

**1.**d; **2.**d; **3.**b; **4.**c; **5.**c; **6.**b; **7.**b; **8.**d; **9.**c; **10.**b; **11.**a; **12.**a; **13.**d; **14.**a; **15.**b; **16.**d; **17.**b; **18.**a; **19.**d; **20.**c; **21.**d; **22.**c; **23.**b; **24.**a; **25.**d

# 3

# Structure and Function of Cells

## Chapter Summary and Key Concepts

*After reading and studying this chapter you should know the following:*

**Sections 3.1, 3.2, 3.3, 3.4**

1. The cell doctrine states that all living things are composed of cells, a cell is the smallest unit of life, and all cells come from preexisting cells.

2. Eukaryotic cells have a membrane-bound nucleus; prokaryotic cells do not.

3. Eukaryotic cells have a plasma membrane, a nucleus, and cytoplasm.

4. Cells remain small to stay efficient.

5. The plasma membrane is selectively permeable and consists of a lipid bilayer, proteins, cholesterol, and carbohydrates that extend from the outer surface.

6. Passive transport moves molecules with the gradient and requires no energy; active transport moves molecules against the gradient and requires energy.

7. Membrane proteins are required for some processes of diffusion, facilitated diffusion, and active transport.

8. Osmosis is the net diffusion of water across a selectively permeable membrane.

9. Endocytosis and exocytosis move materials in bulk. Endocytosis moves material into the cell, and exocytosis moves material out of the cell.

10. Receptor proteins receive and transmit information across the plasma membrane.

11. The sodium-potassium pump is an important active transport system that helps to maintain ion gradients and cell volume.

**Sections 3.5, 3.6**

12. The nucleus is surrounded by a nuclear membrane consisting of two lipid bilayers studded with pores.

13. Ribosomes synthesize proteins and may be free in the cytoplasm or bound to the rough endoplasmic reticulum.

14. The smooth endoplasmic reticulum synthesizes nonprotein molecules and packages proteins in vesicles for movement to the Golgi apparatus.

15. The Golgi apparatus refines proteins and packages them to be shipped.

16. Vesicles are membrane-bound sacs that enclose substances within the cell. They may be secretory, endocytotic, for shipping and storage, or lysosomal.

17. Mitochondria harvest the energy from ingested nutrients and use it to synthesize ATP.

18. Cell structures that contain microtubules include the cytoskeleton which provides structural support, cilia and flagella that aid movement, and centrioles that participate in cell division.

**Section 3.7**

19. Cells use and transform matter and energy through chemical reactions. Metabolism is the sum of all chemical reactions in an organism.

20. Cellular respiration breaks down glucose in the presence of oxygen to yield 36 molecules of ATP. In the absence of oxygen, a cell can produce two molecules of ATP using anaerobic pathways.

# Exercises

*Complete the exercises for each section after you have read and studied the section. If you cannot answer some questions, or answer them incorrectly, return to the chapter and review this information. You may find it helpful to work on only one section at a time. When you have completed all sections, take the Chapter Test as an indicator of your mastery of this topic.*

**3.1   Cells are classified according to their internal organization**

**3.2   Cell structure reflects cell function**

**3.3   A plasma membrane surrounds the cell**

**3.4   Molecules cross the plasma membrane in several ways**

## Matching

____ 1. **cell doctrine**  a. the movement of molecules from one area to another as a result of random motion

____ 2. **plasma membrane**  b. material within a cell, composed of fluid and organelles

____ 3. **eukaryotes**  c. a solution with a concentration of solutes higher than the intracellular fluid

____ 4. **nucleus**  d. diffusion of substances through the plasma membrane with the help of transport proteins

____ 5. **cytoplasm**  e. three general principles that describe cells and their relationship to living things

____ 6. **organelle**  f. movement of a substance through the plasma membrane against the concentration gradient

____ 7. **prokaryotes**  g. movement of material out of the cell by fusion of a vesicle with the plasma membrane

____ 8. **passive transport**  h. the outer membrane that surrounds all cells

____ 9. **diffusion**  i. the nature of the plasma membrane to allow some substances to cross by diffusion, but not others

____ 10. **selective permeability**  j. an active transport pump important in maintaining cell volume

____ 11. **osmosis**  k. movement of materials into the cell by enclosure in a vesicle formed by the plasma membrane

____ 12. **facilitated transport**  l. membrane proteins with specific receptor sites that can trigger biochemical events within the cell

____ 13. **transport protein**  m. organisms composed of cells that have a membrane-bound nucleus

____ 14. **active transport**  n. a solution with the same solute concentration as the intracellular fluid

____ 15. **sodium-potassium pump**  o. the net diffusion of water across a selectively permeable membrane

____ 16. **exocytosis**  p. transport of a molecule through the plasma membrane without the use of energy

____ 17. **endocytosis**  q. organisms whose cells do not have a membrane-bound nucleus

____ 18. **vesicle**  r. a solution with a concentration of solutes lower than the intracellular fluid

____ 19. **receptor protein**  s. structure within a cell that carries out specialized functions

____ 20. **isotonic solution**  t. membrane bound structure that contains a cell's genetic material

____ 21. **hypertonic solution**  u. membrane bound structure that transports substances into, out of, and within a cell

____ 22. **hypotonic solution**  v. required for some membrane transport processes to move a substance across the plasma membrane

## Fill-in-the-Blank

*Referenced sections are indicated in parentheses.*

23. Eukaryotic cells contain three basic structural components, the _____, _____, and _____ _____. (3.1)

24. Some cells increase their surface area without significantly increasing their volume with a projection of the cell membrane called _____. (3.2)

25. The _____ _____ microscope bombards a sample with a beam of electrons to produce a two-dimensional image. (3.2)

26. The plasma membrane is composed of two layers of phospholipids called the _____ _____. (3.3)

27. The changing structure and protein pattern of the plasma membrane is referred to by the term _____ _____. (3.3)

28. A _____ exists when there is a difference in the concentration of a solution in different areas. (3.4)

29. _____ exists when molecules are diffusing randomly, but equally, in all directions. (3.4)

30. The fluid pressure required to oppose osmosis is called _____ _____. (3.4)

31. When ATP is used to power active transport, it will break down to produce _____, _____, and _____. (3.4)

32. The area outside a cell is referred to as the _____ environment. (3.4)

33. When the sodium-potassium pump is being used to reduce cell volume, the _____ of the pump will increase. (3.4)

## Diagram

34. Draw a section of a plasma membrane, labeling each component. You should include the following components in your diagram: a. phospholipid, b. phospholipids bilayer, c. cholesterol, d. glycoprotein, e. receptor protein, f. transport protein, g. open protein channel, h. gated protein, i. carbohydrate groups, j. cytoskeleton filaments.

## Completion

*For each of the membrane transport processes listed below, indicate **all** of the following characteristics that apply in the table on page 35 (characteristics may be used more than once).*

a. requires energy

b. does not require energy

c. moves a substance through the lipid bilayer of the plasma membrane

d. moves a substance through a protein channel

e. moves a substance with the help of a transport protein

f. moves a substance with the gradient

g. moves a substance against the gradient

h. moves a substance within a vesicle

i. moves water only

| Transport Process | Characteristics |
|---|---|
| 35. Diffusion through the lipid bilayer | |
| 36. Diffusion through protein channels | |
| 37. Facilitated transport | |
| 38. Osmosis | |
| 39. Active transport | |
| 40. Endocytosis | |
| 41. Exocytosis | |

## Word Choice

*Circle the accurate term or phrase to complete each sentence.*

42. Cells must maintain a (small/large) size to be efficient. (3.2)

43. The concentration of water in a solution is (the same as/opposite to) the concentration of solutes in the solution. (3.4)

44. Polar molecules (can/cannot) cross the lipid bilayer portion of the cell membrane. (3.4)

45. When a substance moves against the gradient, it moves from an area of (low/high) concentration to an area of (low/high) concentration. (3.4)

46. Gated protein channels are normally (open/closed) and will (open/close) under certain conditions. (3.4)

47. Moving a substance with the gradient (does/does not) require energy. (3.4)

48. The sodium-potassium pump moves three sodium ions (into/out of) the cell and two potassium ions (into/out of) the cell. (3.4)

49. In a hypertonic environment, water will move (into/out of) the cell. (3.4)

# Study Guide for Human Biology

3.5 Internal structures carry out specific functions

3.6 Cells have structures for support and movement

## Crossword Puzzle

### Across

1. Synthesizes ATP
4. Vesicles containing enzymes that detoxify wastes
9. Rodlike structures composed of microtubules that aid the process of cell division
10. An organelle that contains the cell's genetic material
12. The _____ endoplasmic reticulum synthesizes lipids
14. The endoplasmic _____ is a membranous tunnel-like system running through the cytoplasm

### Down

2. A network of microtubules and microfilaments that provide structural support to the cell
3. The _____ apparatus refines proteins and packages them to be shipped to their final destination
5. Responsible for protein synthesis in the cytoplasm
6. Vesicles containing powerful digestive enzymes
7. The _____ endoplasmic reticulum has attached ribosomes
8. Long whiplike projection, found in humans only on sperm cells
11. A membrane-bound sphere that encloses something within the cell
13. Short hairlike projections found covering a cell surface

## Labeling

*Label the indicated structures in Figure 3.2.*

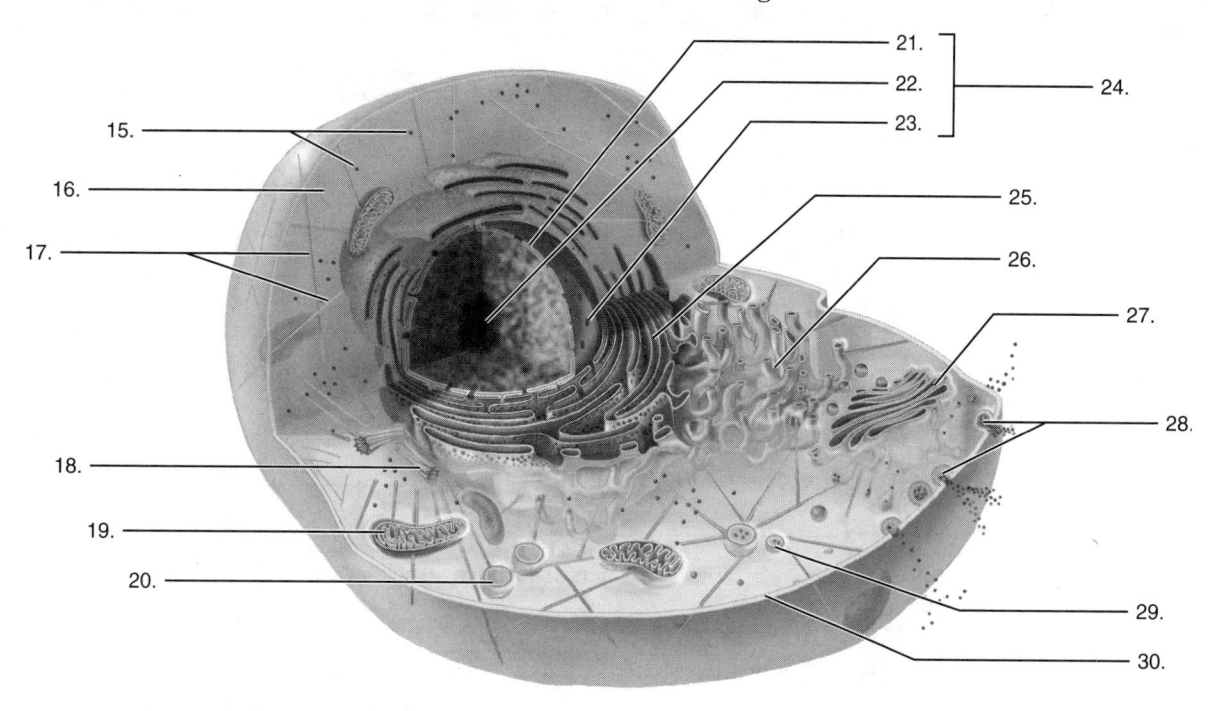

**Figure 3.2**

### 3.7 Cells use and transform matter and energy

## Matching

_____ 1. **metabolism**  a. electron transport molecules

_____ 2. **metabolic pathway**  b. the starting material in a chemical reaction

_____ 3. **substrate**  c. a metabolic pathway in which large molecules are broken down into smaller molecules

_____ 4. **product**  d. production of ATP in the absence of oxygen

_____ 5. **anabolism**  e. the sum of all the chemical reactions in an organism

_____ 6. **catabolism**  f. metabolic processes within a cell that utilize oxygen and produce carbon dioxide in the process of making ATP

_____ 7. **cellular respiration**  g. the ending material in a chemical reaction

_____ 8. **NAD⁺, FAD**  h. a metabolic pathway in which small molecules are assembled into larger molecules

_____ 9. **anaerobic pathway**  i. a series of reactions following one another in an orderly and predictable manner.

## Paragraph Completion

*Use the following terms to complete the paragraphs, and then repeat this exercise with the terms covered. Terms used more than once are listed multiple times.*

| | | | |
|---|---|---|---|
| hydrogen | acetyl groups | FADH$_2$ | cytoplasm |
| ATP synthase | carbohydrates | pyruvate | lipids |
| mitochondria | two | coenzyme A | proteins |
| acetyl CoA | inner | hydrogen | NADH |
| cyclic | gradient | energy | two |
| active transport | inner compartment | oxidative phosphorylation | pyruvate |

The production of ATP occurs primarily in the (10) _____ and involves the catabolism of (11) _____, (12) _____, and (13) _____. Cellular respiration often begins with a molecule of glucose and proceeds through the following stages:

**Glycolysis:** Glycolysis occurs in the (14) _____. During glycolysis, a 6-C glucose molecule is broken down into two 3-C molecules called (15) _____. A net yield of (16) _____ ATP occurs.

**Preparatory Step:** In this preparatory step for the citric acid cycle, the two (17) _____ molecules produced in glycolysis enter the mitochondria and are converted into two molecules of (18) _____. Only the (19) _____ actually enter the citric acid cycle; (20) _____ acts as a shuttle.

**Citric Acid Cycle:** The citric acid cycle is an example of a (21) _____ pathway. During these chemical reactions, the two carbon acetyl molecules are broken down and (22) _____ ions and electrons are transferred to NAD$^+$ and FAD, forming (23) _____ and (24) _____. (25) _____ molecules of ATP are also generated by the citric acid cycle; however, most of the energy of glucose is still trapped in the (26) _____ being carried by NADH and FADH$_2$.

**Electron Transport System:** The electron transport system is located on the (27) _____ membrane of the mitochondria. Hydrogen electrons delivered by NADH and FADH$_2$ are accepted by the proteins of the electron transport system, and release (28) _____ as they move from one transport molecule to the next. This energy is used for the (29) _____ of H⁺ into the outer compartment, creating a H⁺ (30) _____. H⁺ are allowed to flow back into the (31) _____ only through a special protein channel that functions as an enzyme called (32) _____. This enzyme uses the energy released by the flowing H⁺ to synthesize ATP. The phosphorylation of ADP to produce ATP with energy that involves electron transfer is called (33) _____.

## Short Answer

34. What happens to pyruvate when oxygen is not available to the cell?

35. How does substrate level phosphorylation differ from oxidative phosphorylation?

36. What is the final fate of low energy H⁺ and electrons?

37. Why do fats store more than twice the energy of carbohydrates?

# Chapter Test

## Multiple Choice

1. Which of the following is not part of the cell doctrine?
    a. All living things are composed of cells.
    b. Cells must remain small to function effectively.
    c. Cells are the smallest units of life.
    d. All cells come from preexisting cells.

2. You suspect a cell to be prokaryotic. Which of the following would indicate you are mistaken?
    a. The cell has a plasma membrane.
    b. The cell has a nuclear membrane.
    c. The cell has genetic material.
    d. The cell has cytoplasm.

3. The plasma membrane is composed of lipids arranged in a:
   a. single layer with the hydrophilic heads near the extracellular fluid.
   b. single layer with the hydrophobic tails near the extracellular fluid.
   c. bilayer with the tails facing the intracellular and extracellular fluid.
   d. bilayer with the heads facing the intracellular and extracellular fluid.

4. Cholesterol in the plasma membrane functions:
   a. as a receptor molecule.
   b. as a transport molecule.
   c. to make the membrane more rigid.
   d. to make the membrane more permeable.

5. Cells with microvilli:
   a. have less surface area than other cells.
   b. have a surface covered with small projections of the plasma membrane.
   c. have multiple nuclei.
   d. are associated with transmitting information between cells.
   e. b and d.

6. The "fluid mosaic" model of the plasma membrane refers to:
   a. the ability of proteins to drift about in the membrane.
   b. the fluidity of the lipid bilayer.
   c. the presence of organelles in the cytoplasm.
   d. a and b.
   e. all of the above.

7. Passive transport:
   a. requires energy.
   b. relies on the process of diffusion.
   c. is dependent on active transport.
   d. moves a substance against the gradient.

8. Which of the following move bulk substances into the cell by forming a membrane-bound vesicle?
   a. passive transport
   b. osmosis
   c. endocytosis
   d. exocytosis

9. Oxygen and carbon dioxide move into and out of the cell by:
   a. diffusion through the lipid bilayer.
   b. facilitated diffusion.
   c. active transport.
   d. osmosis.

10. Substance X is more concentrated in the intracellular fluid than the extracellular fluid. Substance X will move into the cell by:
    a. diffusion.
    b. facilitated diffusion.
    c. exocytosis.
    d. active transport.

11. How does facilitated diffusion differ from active transport?
    a. Facilitated diffusion requires transport proteins.
    b. Facilitated diffusion does not require energy.
    c. Facilitated diffusion moves a substance down the concentration gradient.
    d. Both b and c.
    e. All of the above.

12. The sodium-potassium pump is important in maintaining cell volume because:
    a. it transports water into and out of the cell.
    b. it blocks the movement of water with membrane proteins.
    c. it transports ions, which results in the movement of water by osmosis.
    d. it counteracts the effect of osmotic pressure.

13. Which of the following events may contribute to a transfer of information across the plasma membrane?
    a. receptor proteins receive and transmit information
    b. a molecule binds a receptor site on a receptor protein
    c. endocytosis brings receptor proteins into the cell
    d. both a and b
    e. all of the above

14. A cell is placed in a hypotonic environment. Which of the following will occur?
    a. Water will move into the cell and the cell will swell.
    b. Water will move out of the cell and the cell will shrink.
    c. Solutes will move into the cell and the cell will shrink.
    d. Solutes will move out of the cell and the cell will swell.

15. The nucleolus is:
    a. an area of concentrated DNA.
    b. an area of RNA synthesis.
    c. an area outside the nucleus where ribosomes are assembled.
    d. a small additional nucleus found in large cells.

16. A cell that lacked smooth endoplasmic reticulum (ER) would be unable to:
    a. synthesize lipids.
    b. detoxify wastes.
    c. manufacture ATP.
    d. synthesize proteins for use outside the cell.

17. Vesicles fusing with the Golgi apparatus would have arisen from the:
    a. plasma membrane.
    b. smooth ER.
    c. mitochondria.
    d. nucleus.

18. Vesicles fusing with the plasma membrane in the process of _____ would have arisen from the _____.
    a. exocytosis, Golgi apparatus
    b. endocytosis, Golgi apparatus
    c. exocytosis, rough ER
    d. endocytosis, nucleus

19. Cellular debris will be removed by:
    a. peroxisomes.
    b. endocytotic vesicles.
    c. secretory vesicles.
    d. lysosomes.

20. Which of the following is composed of a circle of nine microtubule pairs surrounding a single pair in the center?
    a. centrioles
    b. cilia
    c. mitochondria
    d. ribosomes

21. Glycolysis is all of the following *except:*
    a. aerobic.
    b. a linear pathway.
    c. a process that occurs in the cytoplasm.
    d. a process that produces ATP.

22. The breakdown of macromolecules is a(n) _____ process and _____ energy.
    a. anabolic, requires
    b. anabolic, releases
    c. catabolic, requires
    d. catabolic, releases

23. Which of the following stages of cellular respiration does not contain any carbon atoms from the original glucose molecule?
    a. citric acid cycle
    b. electron transport system
    c. glycolysis
    d. formation of acetyl CoA

24. How much ATP could a cell produce if no molecules of NAD$^+$ and FAD were available?
    a. 0
    b. 4
    c. 12
    d. 32

25. What would happen to ATP production in the mitochondria if the inner mitochondrial membrane was freely permeable to H$^+$?
    a. ATP synthesis would proceed with no change.
    b. ATP synthesis would be slightly reduced.
    c. ATP synthesis would be slightly increased.
    d. ATP synthesis would stop.

# Key Concept Review Questions

*Each of the Key Concepts listed at the beginning of this chapter has been rewritten as a question below. After successively completing the study guide exercises and the Chapter Test, you should be able to answer each of these questions. Refer to the Key Concepts list at the beginning of this chapter to check your answers.*

1. What three principles are stated by the cell doctrine?
2. How are eukaryotic cells different from prokaryotic cells?
3. What three basic components are found in eukaryotic cells?
4. Why are all cells small?
5. How permeable is the plasma membrane? What components are found as part of the plasma membrane?
6. How does passive transport differ from active transport?
7. What transport processes require membrane proteins?

8. What is osmosis?
9. What occurs in endocytosis and exocytosis?
10. What is the role of receptor proteins?
11. Why is the sodium-potassium pump important to the cell?
12. How would you describe the structure of the nuclear membrane?
13. What is the function of ribosomes? Where are ribosomes located?
14. What is the function of the smooth endoplasmic reticulum?
15. What is the function of the Golgi apparatus?
16. What are vesicles? List four different types of vesicles.
17. What is the function of mitochondria?
18. List four cell structures that contain microtubules, and describe the function of each.
19. How do cells use and transform energy? What is metabolism?
20. What molecules are involved in the process of cellular respiration? How much ATP is produced? How is this process different in the absence of oxygen?

# Answer Key

## Sections 3.1, 3.2, 3.3, 3.4

**1.**e; **2.**h; **3.**m; **4.**t; **5.**b; **6.**s; **7.**q; **8.**p; **9.**a; **10.**i; **11.**o; **12.**d; **13.**v; **14.**f; **15.**j; **16.**g; **17.**k; **18.**u; **19.**l; **20.**n; **21.**c; **22.**r; **23.** nucleus, cytoplasm, plasma membrane; **24.** microvilli; **25.** transmission electron; **26.** lipid bilayer; **27.** fluid mosaic; **28.** gradient; **29.** Equilibrium; **30.** osmotic pressure; **31.** ADP, P, energy; **32.** extracellular; **33.** activity; **34.** Refer to Fig. 3.5 in your text.; **35.**b,c; **36.**b,d,f; **37.**b,e,f; **38.**b,e,d,f; **39.**a,e,g; **40.**a,h; **41.**a,h; **42.** small; **43.** opposite to; **44.** cannot; **45.** low, high; **46.** closed, open; **47.** does not; **48.** out of, into; **49.** out of

## Sections 3.5, 3.6

**1.** mitochondria; **2.** cytoskeleton; **3.** Golgi; **4.** peroxisomes; **5.** ribosomes; **6.** lysosomes; **7.** rough; **8.** flagellum; **9.** centrioles; **10.** nucleus; **11.** vesicle; **12.** smooth; **13.** cilia; **14.** reticulum; **15.** ribosome; **16.** cytosol; **17.** cytoskeleton; **18.** centriole; **19.** mitochondria; **20.** lysosome; **21.** nuclear membrane; **22.** nucleolus; **23.** nuclear pore; **24.** nucleus; **25.** rough ER; **26.** smooth ER; **27.** Golgi apparatus; **28.** secretory vesicle; **29.** peroxisome; **30.** plasma membrane

## Section 3.7

**1.**e; **2.**i; **3.**b; **4.**g; **5.**h; **6.**c; **7.**f; **8.**a; **9.**d; **10.** mitochondria; **11.** carbohydrates; **12.** lipids; **13.** proteins; **14.** cytoplasm; **15.** pyruvate; **16.** two; **17.** pyruvate; **18.** acetyl CoA; **19.** acetyl groups; **20.** coenzyme A;

**21.** cyclic; **22.** hydrogen; **23.** NADH; **24.** FADH$_2$; **25.** Two; **26.** hydrogen; **27.** inner; **28.** energy; **29.** active transport; **30.** gradient; **31.** inner compartment; **32.** ATP synthase; **33.** oxidative phosphorylation; **34.** It is converted to lactic acid instead of acetyl CoA.; **35.** Oxidative phosphorylation is the only process that requires oxygen and uses energy associated with electron transport.; **36.** They will be accepted by oxygen in the inner compartment to form water molecules.; **37.** Fats contain long chains of carbon atoms that are broken down into many acetyl groups that can enter the Krebs cycle.

## Chapter Test

**1.** b; **2.** b; **3.** d; **4.** c; **5.** b; **6.** d; **7.** b; **8.** c; **9.** a; **10.** d; **11.** d; **12.** c; **13.** d; **14.** a; **15.** b; **16.** a; **17.** b; **18.** a; **19.** d; **20.** b; **21.** a; **22.** d; **23.** b; **24.** a; **25.** d

# 4

# From Cells to Organ Systems

## Chapter Summary and Key Concepts

*After reading and studying this chapter you should know the following:*

**Sections 4.1, 4.2, 4.3**

1. Multicellular organisms consist of many cells, allowing them to achieve greater size and greater ability to maintain a desired environment.

2. Tissues are groups of cells that perform common functions. The four major tissue types of the body are epithelial, connective, muscle, and nervous tissue.

3. Epithelial tissue covers or lines body surfaces or cavities, and is classified by the number of cell layers and cell shape.

4. Glands are a type of epithelial tissue specialized to secrete a product.

5. A basement membrane anchors epithelial tissue to the underlying connective tissue. Junctions anchor epithelial cells to other cells.

6. Connective tissue supports soft organs and connects parts of the body.

7. Connective tissue consists of a nonliving *matrix* of background substance and fibers, and cells responsible for secreting the matrix.

8. Connective tissue can be classified as fibrous or specialized.

9. Fibrous connective tissue includes the following types: loose, dense, elastic, and reticular.

10. Specialized connective tissue includes the following types: cartilage, bone, blood, and adipose tissue.

11. Each type of connective tissue will vary in the type of matrix, cells, and fibers it contains. The structure of the tissue allows it to perform its function.

### Sections 4.4, 4.5

12. Muscle tissue consists of cells specialized to produce movement. Muscle tissue types include skeletal, smooth, and cardiac.

13. Skeletal muscle cells are voluntary, multinucleated, and attached to bones.

14. Smooth muscle cells are involuntary, have a single nucleus, and are found surrounding hollow organs and tubes.

15. Cardiac muscle cells are involuntary, have a single nucleus, and are found in the heart.

16. Nervous tissue consists of cells specialized to generate and transmit electrical impulses used for communication in the body. Nervous tissue cells include neurons responsible for communication, and glial cells that support the neurons.

### Sections 4.6, 4.7, 4.8

17. Organ systems are composed of groups of organs that perform a broad function. The human body has 11 organ systems.

18. Tissue membranes line body cavities. The four major types of tissue membranes are serous, mucous, synovial, and cutaneous.

19. The integumentary system consists of the epidermis and the dermis, and it functions in protection from dehydration, injury, and invasion; regulation of body temperature; synthesis; and sensation.

20. Multicellular organisms maintain homeostasis using negative feedback control systems. These systems include a controlled variable, a sensor, a control center, and an effector.

# Exercises

*Complete the exercises for each section after you have read and studied the section. If you cannot answer some questions, or answer them incorrectly, return to the chapter and review this information. You may find it helpful to work on only one section at a time. When you have completed all sections, take the Chapter Test as an indicator of your mastery of this topic.*

**4.1  Tissues are groups of cells with a common function**

**4.2  Epithelial tissues cover body cavities and surfaces**

**4.3  Connective tissue supports and connects body parts**

## Matching

___ 1. **tissue**
___ 2. **epithelial tissue**
___ 3. **glands**
___ 4. **exocrine gland**
___ 5. **endocrine gland**
___ 6. **basement membrane**
___ 7. **connective tissue**
___ 8. **collagen fibers**
___ 9. **elastic fibers**
___ 10. **reticular fibers**
___ 11. **fibroblasts**
___ 12. **cartilage**
___ 13. **bone**
___ 14. **blood**
___ 15. **adipose tissue**
___ 16. **adipocytes**

a. a connective tissue composed primarily of collagen fibers, with chondroblasts in lacunae
b. fat cells
c. connective tissue protein fibers that allow a tissue to stretch
d. a connective tissue specialized for fat storage
e. a gland that secretes its product into a hollow organ or duct
f. a group of specialized cells that perform a common function
g. a connective tissue consisting of cells suspended in a fluid called plasma
h. sheets of cells that line or cover body surfaces or cavities
i. connective tissue cells that secrete protein fibers
j. a gland that secretes hormones into the bloodstream
k. a tissue that supports and connects parts of the body
l. epithelial tissue specialized to secrete a product
m. connective tissue fibers that support or provide an internal framework for a body organ
n. connective tissue protein fibers that confer strength to a tissue
o. a supporting layer beneath epithelial tissue that anchors it to connective tissue
p. a hard connective tissue containing calcium and phosphate

Chapter 4 From Cells to Organ Systems  49

## Labeling

*Label each epithelial tissue in Figure 4.1 with the (a) tissue type and (b) function.*

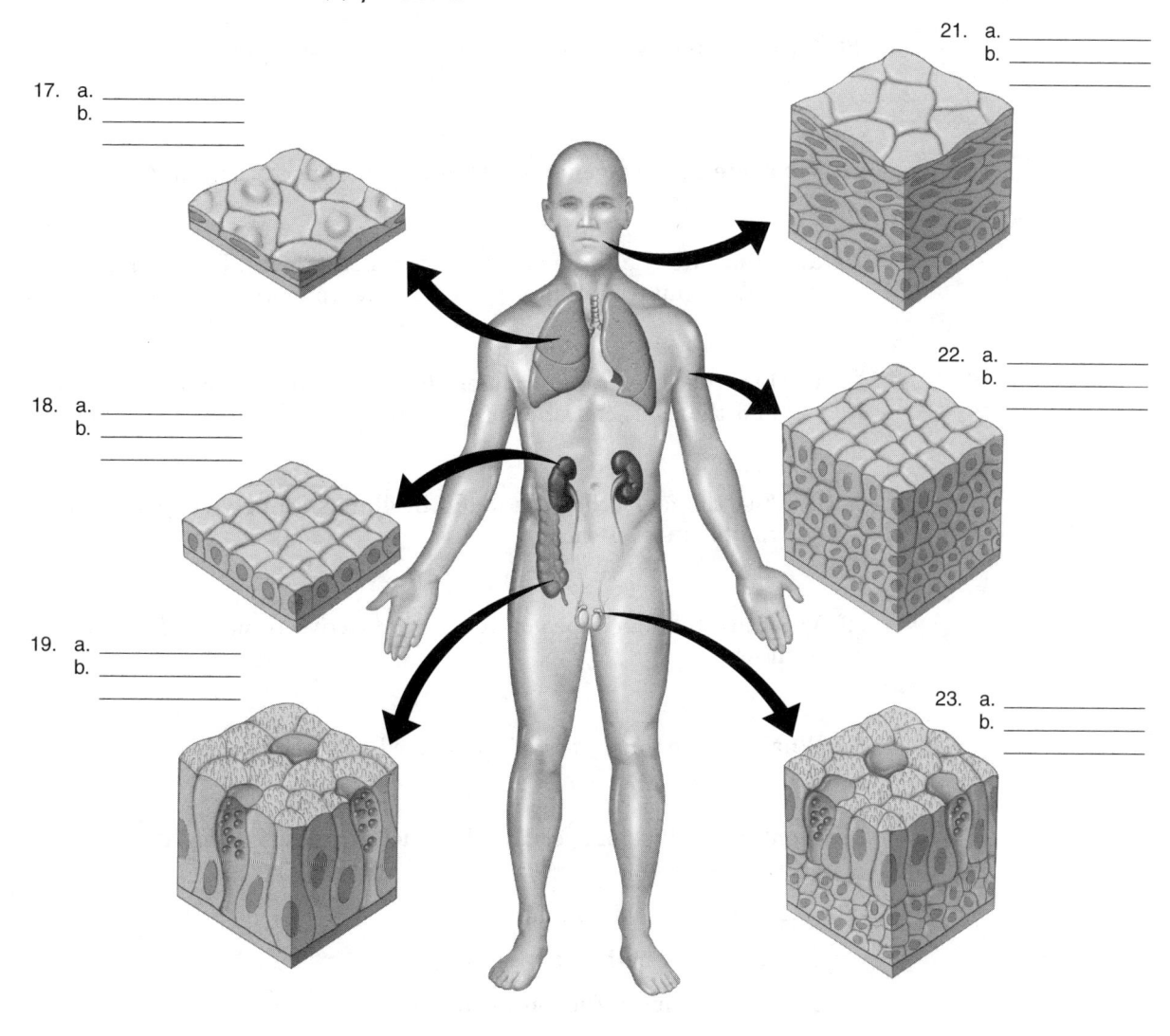

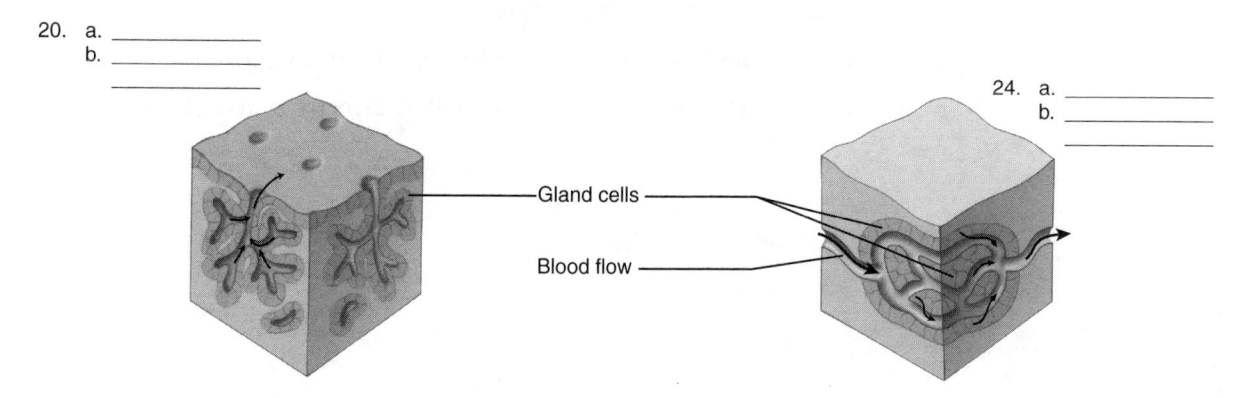

17. a. _____
    b. _____
       _____

18. a. _____
    b. _____
       _____

19. a. _____
    b. _____
       _____

20. a. _____
    b. _____
       _____

21. a. _____
    b. _____
       _____

22. a. _____
    b. _____
       _____

23. a. _____
    b. _____
       _____

24. a. _____
    b. _____
       _____

Gland cells

Blood flow

**Figure 4.1**

## Short Answer

25. List four functions performed by connective tissues.

26. What are the two components found in most types of connective tissue?

27. What are the two general categories of connective tissue?

28. a. What are the three general components found in the extra-cellular matrix of all types of fibrous connective tissue?

    b. List the three types of fibers found in fibrous connective tissue, and the characteristics of each fiber.

    c. List the four cell types found in fibrous connective tissue, and the function of each.

29. What are the four types of fibrous connective tissue, and where is each found?

30. What are the four types of specialized connective tissue?

31. Identify the type of specialized connective tissue in each description below:
    _____ a. has a fluid matrix
    _____ b. has no blood vessels
    _____ c. reduces friction in the joints
    _____ d. has very little ground substance
    _____ e. has a very hard matrix
    _____ f. has mostly matrix with only a few living cells
    _____ g. most of this tissue is located directly under the skin

4.4 **Muscle tissues contract to produce movement**

4.5 **Nervous tissue transmits impulses**

## Matching

___ 1. **muscle tissue**     a. cells of the nervous tissue that support neurons

___ 2. **skeletal muscle tissue**     b. muscle tissue that surrounds hollow organs

___ 3. **cardiac muscle tissue**     c. a tissue specialized for movement

___ 4. **smooth muscle tissue**     d. nervous tissue cells that generate and transmit electrical impulses

___ 5. **nervous tissue**     e. muscle tissue connected to bone by tendons

___ 6. **neurons**     f. muscle tissue found in the heart

___ 7. **glial cells**     g. a tissue specialized for generating and transmitting electrical impulses

## Completion

*Complete the table by filling in the characteristics of each type of muscle tissue.*

| Tissue Type | Location | Number of Nuclei | Voluntary/Involuntary |
| --- | --- | --- | --- |
| 8. Cardiac muscle cells | a. | b. | c. |
| 9. Skeletal muscle cells | a. | b. | c. |
| 10. Smooth muscle cells | a. | b. | c. |

### 4.6 Organs and organ systems perform complex functions

## Fill-in-the-Blank

1. A structure composed of two or more tissue types joined together to perform a function is a(n) _____.

2. A(n) _____ system is a group of organs working together to perform a broad function.

3. A _____ membrane consists of a layer of epithelial tissue and a layer of connective tissue.

4. A general term for a thin layer that covers or surrounds something is _____.

5. _____ membranes line thin cavities between bones in movable joints.

6. Body cavities are lined by _____ membranes.

7. The skin is also referred to as a _____ membrane.

8. _____ membranes line the airways, digestive tract, and reproductive passages.

9. A _____ plane divides the body into top and bottom portions.

10. A structure nearer to a point of reference is described as _____.

11. The thoracic cavity includes the _____ cavities and the _____ cavity.

12. The thoracic and abdominal cavities constitute the _____ cavity.

13. The thoracic and abdominal cavities are separated by the _____.

## Completion

*Fill in the table below by naming each organ system described.*

| Organ System | Description |
| --- | --- |
| 14. | Responsible for gas exchange between the air and the blood. |
| 15. | Produces or resists movement, generates heat. |
| 16. | Transports material through the body, helps maintain body temperature, aids body defenses. |
| 17. | Produces eggs and sperm. |
| 18. | Excretes wastes, maintains volume and composition of body fluids. |
| 19. | Detects stimuli in the external and internal environment and coordinates a response. |

| Organ System | Description |
| --- | --- |
| 20. | Provides the structural framework for movement of the body. |
| 21. | Provides protection from injury, infection, and dehydration; receives sensory input. |
| 22. | Returns excess tissue fluid to the circulatory system and aids body defense mechanisms. |
| 23. | Produces hormones that regulate many body functions. |
| 24. | Provides the body with water and nutrients. |

## 4.7 The skin as an organ system

# Crossword Puzzle

### Across

2. Pigment-producing cells in the epidermis
4. Sweat glands produce sweat that contains this antibiotic secretion
5. The skin represents this organ system
7. A supporting layer for the skin that contains adipose tissue and loose connective tissue
9. Cells at the base of the epidermis that are constantly dividing
11. Small projections on the surface of the dermis
12. A waterproof protein produced by keratinocytes in the epidermis
13. The region of a hair that is found below the skin's surface
14. The region of a hair that extends above the skin's surface

### Down

1. The outer layers of the epidermis are dead because these are not found in the epidermis
3. This type of muscle is found at the base of each hair follicle
6. The outer layer of the skin's epithelial tissue
8. A type of gland that produces oil
10. This type of epithelial cell is found in the epidermis

### 4.8 Multicellular organisms must maintain homeostasis

## Ordering

*Place these events of a negative feedback cycle in the correct sequence; then answer the questions that follow.*

   a. core temperature returns to normal
   b. hypothalamus decreases impulses to the blood vessels and activates the sweat glands
   c. core temperature rises
   d. skin and internal organs sense a change in temperature
   e. blood vessels dilate and perspiration increases
   f. skin and internal organs send a message to the hypothalamus

1. _____
2. _____
3. _____
4. _____
5. _____
6. _____

_____  7. What is the controlled variable?

_____  8. What acts as the sensor?

_____  9. What acts as the control center?

_____  10. What acts as the effector?

_____  11. What type of feedback will amplify a change rather than return body systems to normal?

# Chapter Test

## Multiple Choice

1. When a group of cells perform a common function they become a(n):
   a. organism.
   b. organ.
   c. tissue.
   d. organ system.

2. _____ tissue covers or lines body surfaces.
   a. Connective
   b. Adipose
   c. Loose
   d. Epithelial

3. A simple epithelium would *not* function in:
   a. transporting materials.
   b. absorbing nutrients.
   c. secreting wastes.
   d. providing protection.

4. All of the following are associated with exocrine glands *except*:
   a. saliva.
   b. ducts.
   c. hormones.
   d. sweat glands.

5. A basement membrane consists of _____ secreted by _____.
   a. carbohydrates, fibroblasts
   b. proteins, epithelial cells
   c. proteins, connective tissue cells
   d. fat, adipocytes

6. Matrix is found in all of the following *except*:
   a. adipose tissue.
   b. cartilage.
   c. elastic connective tissue.
   d. muscle.

7. Cardiac muscle cells must be able to communicate rapidly with each other, a process that involves the movement of ions from one cell to another. Which type of junction aids this process in cardiac muscle cells?
   a. tight junctions
   b. communication junctions
   c. adherence junctions
   d. gap junctions

8. A unique feature of connective tissue is that the cells of the tissue secrete the _____.
   a. plasma membrane
   b. junctions
   c. basement membrane
   d. matrix

9. Which of the following connective tissue elements are incorrectly matched with a characteristic?
   a. elastic fibers–allow tissue to stretch
   b. reticular fibers–secrete matrix
   c. collagen fibers–confer strength
   d. adipocytes–store fat

10. Loose connective tissue and dense connective tissue both contain collagen fibers. What is different about the collagen fibers in these two tissues?
    a. the chemical composition of the fibers
    b. the orientation of the fibers
    c. the collagen in dense connective tissue is more flexible than the collagen in loose connective tissue
    d. the collagen fibers in dense connective tissue are repairable, while the collagen fibers in loose connective tissue are not

11. Cartilage is slow to heal because:
    a. it is a connective tissue.
    b. it is not specialized for growth.
    c. it is seldom injured.
    d. it has no blood vessels.

12. The cells of muscle tissue are called:
    a. adipocytes.
    b. fibroblasts.
    c. chondroblasts.
    d. fibers.

13. The walls of blood vessels contain _____ muscle that is _____.
    a. skeletal muscle, voluntary
    b. skeletal muscle, involuntary
    c. smooth muscle, voluntary
    d. smooth muscle, involuntary

14. In a neuron the _____ receives signals, and the _____ transmits signals.
    a. axon, cell body
    b. cell body, dendrite
    c. dendrite, axon
    d. axon, dendrite

15. The posterior cavity consists of:
    a. the thoracic cavity and the spinal cavity.
    b. the cranial cavity and the thoracic cavity.
    c. the thoracic cavity and the abdominal cavity.
    d. the cranial cavity and the spinal cavity.

16. Tissue membranes consist of:
    a. epithelial tissue and connective tissue.
    b. epithelial tissue and a basement membrane.
    c. a basement membrane and a cellular membrane.
    d. a pleural membrane and a parietal membrane.

17. Goblet cells are found in:
    a. some epithelial tissue.
    b. connective tissue.
    c. mucous membranes.
    d. the hypodermis.

18. Which of the following systems produces hormones that regulate many body functions?
    a. integumentary
    b. nervous
    c. endocrine
    d. reproductive

19. Which of the following terms is incorrectly defined?
    a. proximal—farther away
    b. anterior—nearer the front
    c. transverse plane—divides the body into right and left sides
    d. a and c

20. The hypodermis:
    a. is the middle layer of the skin
    b. provides rigid support for the epidermis
    c. contains fat cells
    d. a and d
    e. b and c

21. Skin pigmentation results from the activity of:
    a. keratinocytes.
    b. basal cells.
    c. fibroblasts.
    d. melanocytes.

22. Which of the following is a false statement about interstitial fluid?
    a. The composition of the interstitial fluid should change regularly.
    b. Interstitial fluid is found between the cells of an organism.
    c. Interstitial fluid makes up the internal environment of an organism.
    d. Cells obtain nutrients from the interstitial fluid.

23. A drop in blood pressure results in constriction of the blood vessels. Blood pressure functions as the _____ in a negative feedback system.
   a. sensor
   b. effector
   c. control center
   d. controlled variable

24. As humans age the skin begins to sag and wrinkle. This is most likely due to a change in:
   a. the cellular junctions.
   b. the blood supply to the skin.
   c. the type of collagen fibers produced.
   d. the amount of skeletal muscle present.

25. Contractions of the uterus during childbirth are an example of positive feedback because:
   a. they are a short-term event.
   b. they are not required to maintain health.
   c. a change in a controlled variable is not monitored by sensors.
   d. a change in a controlled variable serves to amplify the change.

# Key Concept Review Questions

*Each of the Key Concepts listed at the beginning of this chapter has been rewritten as a question below. After successfully completing the study guide exercises and the Chapter Test, you should be able to answer each of these questions. Refer to the Key Concepts list at the beginning of this chapter to check your answers.*

1. What are some advantages of being multicellular?
2. What is a tissue? What are the four major tissue types in the body?
3. Where are epithelial tissues found, and how are they classified?
4. What are glands? What is their function?
5. What is the function of a basement membrane? What is the function of a cellular junction?
6. What is the broad function of connective tissue?
7. What components make up connective tissue?
8. What are the two broad categories of connective tissue?
9. What are four types of fibrous connective tissue?
10. What are four types of specialized connective tissue?
11. How are connective tissue types different?

12. What is the function of muscle tissue? List three types of muscle tissue.

13. What are three characteristics of skeletal muscle cells?

14. What are three characteristics of smooth muscle cells?

15. What are three characteristics of cardiac muscle cells?

16. What is the function of nervous tissue? Name two types of nervous tissue cells and their function.

17. What is an organ system? How many organ systems are found in the human body?

18. What is the function of a tissue membrane? Name the four major types of tissue membranes.

19. What are the two main components of the integumentary system? What are the functions of this system?

20. How do multicellular organisms maintain homeostasis? What is involved in this mechanism?

---

# Answer Key

## Sections 4.1, 4.2, 4.3

**1.** f; **2.** h; **3.** l; **4.** e; **5.** j; **6.** o; **7.** k; **8.** n; **9.** c; **10.** m; **11.** i; **12.** a; **13.** p; **14.** g; **15.** d; **16.** b; **17.** a. simple squamous, b. permits exchange of nutrients, wastes, and gases; **18.** a. simple cuboidal; b. secretes and reabsorbs water and small molecules; **19.** a. simple columnar; b. absorbs nutrients and produces mucus; **20.** a. exocrine gland, b. secretes product into a hollow organ or duct; **21.** a. stratified squamous; b. protection; **22.** a. stratified cuboidal, b. secretes water and ions; **23.** a. stratified columnar, b. secretes mucus; **24.** a. endocrine gland, b. secretes hormones into the bloodstream; **25.** support internal organs, connect structures within the body, store fat, produce blood cells; **26.** living cells, nonliving extracellular matrix; **27.** fibrous, specialized; **28.** a. fibers, cells, ground substance, b. collagen fibers (strength), elastic fibers (flexibility without breaking), reticular fibers (form a mesh-like supportive framework); c. fat cells (store fat), mast cells (aid the immune system), white blood cells (aid the immune system), fibroblasts (secrete the fibers of the matrix); **29.** loose (surrounds organs and blood vessels), dense (tendons, ligaments, skin), elastic (surrounds organs that change shape), reticular (provides an internal framework for soft organs); **30.** cartilage, bone, blood, adipose tissue; **31.** a. blood, b. cartilage, c. cartilage, d. adipose tissue, e. bone, f. bone, g. adipose tissue

## Sections 4.4, 4.5

**1.** c; **2.** e; **3.** f; **4.** b; **5.** g; **6.** d; **7.** a; **8.** a. heart, b. one, c. involuntary; **9.** a. attached to bones, b. multiple, c. voluntary; **10.** a. hollow organs and tubes, b. one, c. involuntary

## Section 4.6

1. organ; 2. organ; 3. tissue; 4. membrane; 5. Synovial; 6. serous; 7. cutaneous; 8. Mucous; 9. transverse; 10. proximal; 11. pleural, pericardial; 12. anterior; 13. diaphragm; 14. respiratory; 15. muscular; 16. circulatory; 17. reproductive; 18. urinary; 19. nervous; 20. skeletal; 21. integumentary; 22. lymphatic; 23. endocrine; 24. digestive

## Section 4.7

1. blood vessels; 2. melanocytes; 3. smooth; 4. dermicidin; 5. integumentary; 6. epidermis; 7. hypodermis; 8. sebaceous; 9. basal; 10. squamous; 11. papillae; 12. keratin; 13. root; 14. shaft

## Section 4.8

1. c; 2. d; 3. f; 4. b; 5. e; 6. a; 7. core temperature; 8. skin, internal organs; 9. hypothalamus; 10. blood vessels, sweat glands; 11. positive

## Chapter Test

1. c; 2. d; 3. d; 4. c; 5. b; 6. d; 7. d; 8. d; 9. b; 10. c; 11. d; 12. d; 13. d; 14. c; 15. d; 16. a; 17. d; 18. c; 19. d; 20. c; 21. d; 22. a; 23. d; 24. c; 25. d

# 5

# The Skeletal System

## Chapter Summary and Key Concepts

*After reading and studying this chapter you should know the following:*

**Sections 5.1, 5.2, 5.3**

1. The skeletal system consists of bones, ligaments, and cartilage.

2. The skeleton provides support, movement, and protection. In addition, the bones provide for formation of blood cells and mineral storage.

3. Compact bone tissue consists of osteons containing osteocytes in lacunae that communicate via canaliculi.

4. Spongy bone tissue is a latticework of trabeculae containing osteocytes in lacunae; the spaces between the trabeculae are filled with red marrow.

5. Ligaments attach bone to bone, and tendons attach muscle to bone.

6. The skeletal system contains three types of cartilage: fibrocartilage, hyaline cartilage, and elastic cartilage.

7. Ossification is the process of developing bone from cartilage, which begins in the embryo.

8. Bones lengthen through adolescence, and undergo replacement, repair, and remodeling throughout adulthood.

9. Bone cells include osteocytes that maintain bone tissue, osteoblasts that build bone tissue, and osteoclasts that dissolve bone tissue.

10. Homeostasis of bone tissue requires a balance in the activity of osteoblasts and osteoclasts.

11. Osteoblast and osteoclast activity is regulated by parathyroid hormone and calcitonin.

**Sections 5.4, 5.5, 5.6**

12. Bones are classified by shape as long, short, flat, and irregular.

13. Bones consist of both compact and spongy bone tissue arranged to meet the functional needs of the bone.

14. The skeleton is organized into the axial skeleton and the appendicular skeleton.
15. The axial skeleton consists of the skull, vertebral column, ribs, and sternum.
16. The appendicular skeleton consists of the pectoral girdle, pelvic girdle, and limbs.
17. Joints are points of contact between bones and are called articulations.
18. Joints are classified in three broad categories based on mobility: fibrous joints are immovable, cartilaginous joints are slightly movable, and synovial joints are freely movable.
19. Synovial joints contain a joint cavity lined with synovial membrane and filled with synovial fluid.
20. Diseases and disorders of the skeletal system include sprains, bursitis, tendinitis, osteoarthritis, and rheumatoid arthritis.

# Exercises

*Complete the exercises for each section after you have read and studied the section. If you cannot answer some questions, or answer them incorrectly, return to the chapter and review this information. You may find it helpful to work on only one section at a time. When you have completed all sections, take the Chapter Test as an indicator of your mastery of this topic.*

5.1   The skeletal system consists of connective tissue

5.2   Bone development begins in the embryo

5.3   Mature bone undergoes remodeling and repair

## Matching

_____ 1. **bone**     a. dense fibrous connective tissue structures that attach bone to bone

_____ 2. **compact bone**     b. a strip of cartilage in the epiphysis of a long bone that allows for bone lengthening

_____ 3. **spongy bone**     c. the hard elements of the skeleton

_____ 4. **osteocytes**     d. a bone condition in which large amounts of bone mass are lost

_____ 5. **osteons**     e. cartilage-forming cells

_____ 6. **central canal**     f. dense bone tissue that forms the shaft of a long bone

_____ 7. **ligaments**     g. bone-dissolving cells

_____ 8. **cartilage**     h. a ringlike arrangement of bone tissue formed in compact bone

___ 9. **chondroblasts**     i. less dense bone tissue that is light but strong

___ 10. **osteoblasts**     j. a passageway through the center of an osteon that carries blood vessels

___ 11. **growth plate**     k. a connective tissue that contributes to the function of the skeletal system by forming embryonic structures and aiding movement

___ 12. **osteoclast**     l. bone-forming cells

___ 13. **osteoporosis**     m. mature bone cells that maintain the structure of bone

## Fill-in-the-Blank

*Referenced sections are in parentheses.*

14. The three functions of the skeleton are _____, _____, and _____. (5.1)

15. Two additional functions of bone are _____ and _____. (5.1)

16. The three types of connective tissue that make up the skeleton are _____, _____, and _____. (5.1)

17. The intervertebral discs are composed of _____. (5.1)

18. _____ cartilage covers the ends of mature bones. (5.1)

19. The most flexible type of cartilage is _____. (5.1)

20. During bone remodeling and repair _____ break down bone tissue and _____ build new bone tissue. (5.3)

21. Electrical currents within bone are caused by _____ and increase the activity of _____. (5.3)

22. A mass of clotted blood produced when a bone fractures is a _____. (5.3)

23. A callus is a _____ bond between broken ends of bone, formed by _____ during bone repair. (5.3)

## Labeling

*Identify each structure indicated in Figure 5.1.*

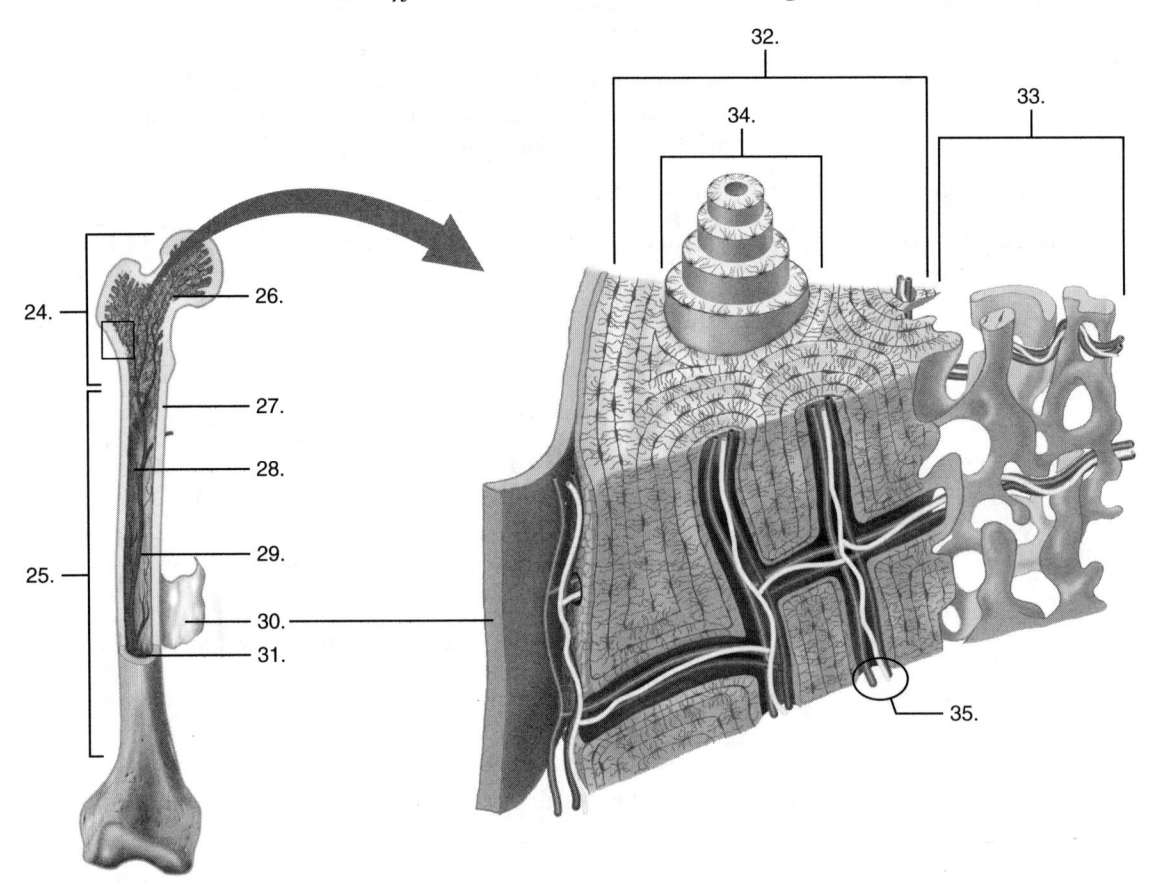

**Figure 5.1**

| 24. _____ | 30. _____ |
| 25. _____ | 31. _____ |
| 26. _____ | 32. _____ |
| 27. _____ | 33. _____ |
| 28. _____ | 34. _____ |
| 29. _____ | 35. _____ |

## Ordering

*Number each of these events, placing them in the proper sequence with the first event numbered "1" and the last event numbered "8."*

36. Embryonic bone development:
    - \_\_\_\_ a. osteoblasts develop into osteocytes
    - \_\_\_\_ b. bone models are constructed of hyaline cartilage
    - \_\_\_\_ c. osteoblasts secrete osteoid
    - \_\_\_\_ d. blood vessels enter the area, bringing osteoblasts
    - \_\_\_\_ e. chondroblasts die
    - \_\_\_\_ f. osteocytes maintain bone matrix
    - \_\_\_\_ g. hydroxyapatite develops in osteoid matrix
    - \_\_\_\_ h. cartilage matrix breaks down

37. Bone repair:
    - \_\_\_\_ a. osteoblasts arrive in the area
    - \_\_\_\_ b. a hematoma develops
    - \_\_\_\_ c. fibroblasts become chondroblasts
    - \_\_\_\_ d. blood vessels bleed into the broken or fractured area
    - \_\_\_\_ e. osteoclasts arrive and destroy bone fragments
    - \_\_\_\_ f. new bone tissue is produced
    - \_\_\_\_ g. a callus forms
    - \_\_\_\_ h. fibroblasts migrate to the injured area

## Word Choice

*Circle the correct term to complete each sentence.*

38. During bone remodeling, new bone is produced in areas of (low/high) compressive stress.

39. Mature compact bone is (one-third/two-thirds) osteoid and (one-third/two-thirds) hydroxyapatite.

40. Once cartilage in the growth plate has been replaced by bone tissue, the bone can continue to grow only in (length/width).

41. Falling levels of calcium in the blood (increase/decrease) the activity of parathyroid hormone.

42. Rising levels of calcium in the blood (increase/decrease) the activity of calcitonin.

43. Calcitonin (increases/decreases) the activity of osteoblasts.

44. Increased osteoclast activity (increases/decreases) the blood level of calcium.

5.4 **The skeleton protects, supports, and permits movement**

5.5 **Joints form connections between bones**

5.6 **Diseases and disorders of the skeletal system**

## Labeling

*Identify each of the bones indicated in Figure 5.2.*

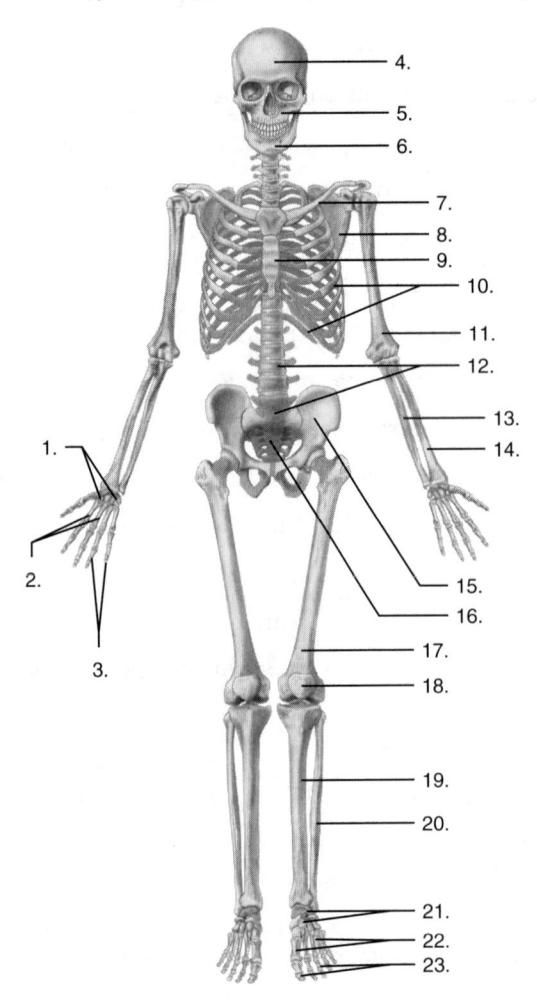

**Figure 5.2**

1. _____
2. _____
3. _____
4. _____
5. _____
6. _____
7. _____
8. _____
9. _____
10. _____
11. _____
12. _____
13. _____
14. _____
15. _____
16. _____
17. _____
18. _____
19. _____
20. _____
21. _____
22. _____
23. _____

## Completion

*Name each injury or disorder of the skeletal system described in the table.*

| Injury or disorder of the skeletal system | Description |
| --- | --- |
| 24. | inflammation of the sinuses |
| 25. | excessive loss of bone mass |
| 26. | swelling and inflammation of tendons, causing pressure on the nerves that supply the hand |
| 27. | joint inflammation caused by a malfunction of the immune system |
| 28. | stretched or torn ligaments accompanied by bruising, swelling, and pain |
| 29. | joint inflammation resulting from an autoimmune malfunction |
| 30. | inflammation of the bursae following injury |

## Crossword Puzzle

**Across**

2. The _____ pubis is a cartilaginous joint between the coxal bones
4. A joint _____ is found only in synovial joints
5. A band of connective tissue that connects bone to bone
7. A point of contact between two vertebrae
9. Cartilage pads that cushion the area between two vertebrae
10. This type of joint is the most freely movable
12. Sacs of fluid that provide cushioning within a joint
13. The _____ girdle supports the upper limbs

**Down**

1. This region of the vertebral column contains seven vertebrae
2. The flexibility of the human skeleton results in a decrease in _____
3. Immovable joints
6. A band of connective tissue that connects muscle to bone
8. This part of the skeleton includes the skull and vertebral column
10. This type of bone is as wide as it is long
11. This region of the vertebral column contains five vertebrae

# Chapter Test

## Multiple Choice

1. Which of the following is not a function of the skeleton?
   a. support
   b. formation of blood cells
   c. regulation of blood cell production
   d. mineral storage

2. In an adult long bone:
   a. the shaft contains yellow marrow.
   b. trabeculae are found in the periosteum.
   c. red marrow produces osteoclasts.
   d. the diaphysis contains the rounded ends of the bone.

3. The periosteum:
   a. lines the inner cavity of a long bone and contains the bone marrow.
   b. covers the ends of a long bone to reduce friction in the joint.
   c. is found at the junction of compact bone and spongy bone.
   d. covers the outer surface of the bone and contains osteoblasts.

4. Osteocytes in compact bone remain in contact with each other via cytoplasmic extensions that extend through:
   a. lacunae.
   b. trabeculae.
   c. canaliculi.
   d. a central canal.

5. Spongy bone and compact bone tissue both consist of a hard mineralized matrix containing calcium and phosphate.
   a. true
   b. false

6. A structure composed of dense fibrous connective tissue that attaches muscle to bone is a:
   a. ligament.
   b. tendon.
   c. osteon.
   d. trabeculae.

7. Hyaline cartilage is found in all of the following *except:*
   a. embryonic structures that will become bone.
   b. the ends of long bones.
   c. cartilaginous joints.
   d. tendons and ligaments.

8. Which of the following occurs first in the process of ossification?
   a. Blood vessels develop in the cartilage model.
   b. Osteoblasts begin building bone tissue.
   c. Chondroblasts die.
   d. Growth plates allow lengthening.

9. A mature bone that had a lower-than-normal percentage of hydroxyapatite would be:
   a. too flexible.
   b. too brittle.
   c. too long.
   d. too short.

10. Replacement of cartilage in the growth plate with bone tissue:
    a. only occurs when a bone has completed all growth and remodeling.
    b. is triggered by rising levels of PTH.
    c. indicates an imbalance in calcitonin and PTH.
    d. occurs when levels of growth hormone are too high.
    e. prevents further lengthening of the bone.

11. What is the role of a mature osteocyte in the body?
    a. to monitor the activity of chondrocytes
    b. to balance the activity of osteoblasts and osteoclasts
    c. to maintain bone tissue
    d. to produce calcitonin and PTH

12. The bones of trained athletes may be thicker and heavier than the bones of nonathletes because:
    a. athletes generally have better nutrition.
    b. athletic activity stimulates higher hormone levels.
    c. bone mass increases when bone is stressed.
    d. athletic activity decreases the activity of osteocytes.

13. Parathyroid hormone stimulates the activity of:
    a. osteocytes.
    b. osteoblasts.
    c. osteoclasts.
    d. chondroblasts.

14. Adequate calcium intake is important in the prevention of osteoporosis. Which of the following might occur if calcium intake was too low?
    a. Low blood calcium levels would stimulate the activity of osteoblasts.
    b. Low blood calcium levels would stimulate the activity of osteoclasts.
    c. Inadequate calcium would be available for deposit into bone tissue.
    d. Osteocytes would die.
    e. b and c

15. The vertebrae are examples of _____ bones.
    a. long
    b. short
    c. flat
    d. irregular

16. Fused vertebrae are found in the _____ region of the vertebral column.
    a. lumbar
    b. coccygeal
    c. thoracic
    d. cervical

17. Cartilage discs between the vertebrae are composed of _____ and function _____.
    a. fibrocartilage, as shock absorbers
    b. hyaline cartilage, as protection
    c. hyaline cartilage, as attachments between vertebrae
    d. elastic cartilage, in flexibility

18. Which of the following is not part of the pelvic girdle?
    a. coxal bone
    b. clavicle
    c. sacrum
    d. coccyx

19. The fingers are composed of:
    a. carpals.
    b. phalanges.
    c. metacarpals.
    d. tarsals.

20. The disadvantage of highly flexible joints is:
    a. inadequate blood supply.
    b. poor control of movement.
    c. poor stability.
    d. limited movement.

21. Which of the following correctly sequences the joints in order of increasing mobility?
    a. synovial, fibrous, cartilaginous
    b. fibrous, synovial, cartilaginous
    c. cartilaginous, fibrous, synovial
    d. fibrous, cartilaginous, synovial

22. The knee joint contains all of the following *except:*
    a. menisci.
    b. bursa.
    c. synovial fluid.
    d. fontanels.

23. A movement that increases the angle of a joint is called:
    a. supination.
    b. rotation.
    c. extension.
    d. circumduction.

24. Inflammation of the joints due to wear and tear is called:
    a. osteoarthritis.
    b. immunoarthritis.
    c. rheumatoid arthritis.
    d. inflammatory arthritis.

25. Which of the following does not apply to sprains?
    a. rapid healing due to a rich blood supply
    b. stretched or torn ligaments
    c. internal bleeding
    d. possible surgery to remove a torn ligament

# Key Concept Review Questions

*Each of the Key Concepts listed at the beginning of this chapter has been rewritten as a question. After successfully completing the study guide exercises and the Chapter Test, you should be able to answer each of these questions. Refer to the Key Concepts list at the beginning of this chapter to check your answers.*

1. What three types of connective tissue are found in the skeletal system?
2. What are three functions of the skeleton and two additional functions of bone?
3. How would you describe the structure of compact bone?
4. How would you describe the structure of spongy bone?
5. What is the function of a ligament? What is the function of a tendon?
6. What three types of cartilage are found in the skeletal system?
7. What is ossification? When does it begin?
8. What bone activities occur in adolescence and in adulthood?
9. What are the names of three types of bone cells and the function of each?
10. What is required for homeostasis of bone tissue?
11. What two hormones regulate the activity of osteoblasts and osteoclasts?
12. What are the four categories of shape used to classify bones?
13. What type of bone tissue is found in each shape of bone?
14. What are the two divisions of the skeleton?
15. What is included in the axial skeleton?
16. What is included in the appendicular skeleton?
17. What is a joint? What is another term for a joint?
18. What are the names of the three categories of joints and the mobility of each?
19. How would you describe the structure of a synovial joint?
20. What are the five disorders of the skeletal system discussed in this chapter?

# Answer Key

## Sections 5.1, 5.2, 5.3

**1.** c; **2.** f; **3.** i; **4.** m; **5.** h; **6.** j; **7.** a; **8.** k; **9.** e; **10.** l; **11.** b; **12.** g; **13.** d; **14.** support, movement, protection; **15.** formation of blood cells, mineral storage; **16.** bone, ligaments, cartilage; **17.** fibrocartilage; **18.** Hyaline; **19.** elastic; **20.** osteoclasts, osteoblasts; **21.** stress, osteoblasts; **22.** hematoma; **23.** fibrocartilage, chondroblasts; **24.** epiphysis; **25.** diaphysis; **26.** spongy bone; **27.** compact bone; **28.** yellow bone marrow; **29.** blood vessel; **30.** periosteum; **31.** central cavity; **32.** compact bone; **33.** spongy bone; **34.** osteon; **35.** blood vessels in central canal; **36.** a.7, b.1, c.5, d.4, e.2, f.8, g.6, h.3; **37.** a.7, b.2, c.4, d.1, e.6, f.8, g.5, h.3; **38.** high; **39.** one-third, two-thirds; **40.** width; **41.** increase; **42.** increase; **43.** increases; **44.** increases

## Sections 5.4, 5.5, 5.6

**1.** carpals; **2.** metacarpals; **3.** phalanges; **4.** cranium; **5.** maxilla; **6.** mandible; **7.** clavicle; **8.** scapula; **9.** sternum; **10.** ribs; **11.** humerus; **12.** vertebrae; **13.** ulna; **14.** radius; **15.** coxal bone; **16.** sacrum; **17.** femur; **18.** patella; **19.** tibia; **20.** fibula; **21.** tarsals; **22.** metatarsals; **23.** phalanges; **24.** sinusitis; **25.** osteoporosis; **26.** carpal tunnel syndrome; **27.** rheumatoid arthritis; **28.** sprain; **29.** osteoarthritis; **30.** bursitis; **Crossword Puzzle: 1.** cervical; **2.** across: symphysis; down: stability; **3.** fibrous; **4.** capsule; **5.** ligament; **6.** tendon; **7.** articulation; **8.** axial; **9.** disk; **10.** across: synovial; down: short; **11.** lumbar; **12.** bursae; **13.** pectoral

## Chapter Test

**1.**c; **2.**a; **3.**d; **4.**c; **5.**a; **6.**b; **7.**d; **8.**c; **9.**a; **10.**e; **11.**c; **12.**c; **13.**c; **14.**e; **15.**d; **16.**b; **17.**a; **18.**b; **19.**b; **20.**c; **21.**d; **22.**d; **23.**c; **24.**a; **25.**a

# 6

# The Muscular System

## Chapter Summary and Key Concepts

*After reading and studying this chapter you should know the following:*

**Sections 6.1, 6.2**

1. Muscles produce movement, generate tension, and generate heat.

2. The three types of muscle are skeletal muscle, cardiac muscle, and smooth muscle.

3. Muscle cells are excitable and act by contracting and then relaxing.

4. Muscles most frequently attach to bone via tendons, but they may also attach to other muscles or to the skin.

5. Muscles have one attachment point called the origin and a second attachment point called the insertion. When muscles contract, the insertion is pulled toward the origin.

6. In skeletal muscle, individual muscle cells are called fibers, and they are arranged in bundles called fascicles. A whole muscle contains many fascicles bound together with connective tissue. The connective tissue layers of a muscle extend beyond the muscle as a tendon.

7. Muscle fibers contain myofibrils, which in turn contain the contractile filaments actin and myosin.

8. The contractile unit of a muscle is the sarcomere.

9. Skeletal muscle cells are stimulated by motor neurons that release the neurotransmitter acetylcholine.

10. During contraction, acetylcholine stimulates the release of calcium ions into the cytoplasm. The presence of calcium triggers the events leading to contraction.

11. Contraction results when actin filaments are pulled across myosin filaments, causing the sarcomere to shorten.

12. Muscle contraction and relaxation require ATP.

### Section 6.3

13. The force of muscle contraction is determined by the number of muscle cells in a motor unit, the number of motor units involved, and the frequency of nerve stimulation.

14. A twitch is a complete cycle of contraction and relaxation. A latent period precedes a twitch.

15. Recruitment is an increase in muscle force due to an increasing number of motor units contracting.

16. Summation is an increase in muscle force due to an increase in the frequency of nerve stimulation.

17. The internal structure and chemistry of a muscle fiber will determine whether it is a slow-twitch fiber or a fast-twitch fiber. Muscle strength and endurance are determined by the number of slow-twitch and fast-twitch fibers present.

### Sections 6.4, 6.5

18. Cardiac muscle cells have one nucleus, intercalated discs, and striations; are involuntary; and are found only in the heart.

19. Smooth muscle cells have one nucleus, gap junctions, and no striations; are involuntary; and are found around hollow organs or tubes.

20. Disorders of the muscular system include muscular dystrophy, tetanus, muscle cramps, pulled muscles, and fasciitis.

## Exercises

*Complete the exercises for each section after you have read and studied the section. If you cannot answer some questions, or answer them incorrectly, return to the chapter and review this information. You may find it helpful to work on only one section at a time. When you have completed all sections, take the Chapter Test as an indicator of your mastery of this topic.*

**6.1 Muscles produce movement or generate tension**

**6.2 Individual muscle cells contract and relax**

### Matching

_____ 1. **skeletal muscle**  a. the contractile unit of a muscle cell

_____ 2. **origin**  b. a process resulting in shortening of a sarcomere

_____ 3. **insertion**  c. a decline in muscle performance during sustained exercise

_____ 4. **myofibrils**  d. muscles that attach to bone

_____ 5. **sarcomere**  e. a chemical messenger molecule that stimulates muscle cells to contract

_____ 6. **myosin**  f. a specialized smooth endoplasmic reticulum that stores calcium ions in muscle cells

_____ 7. **actin**  g. point of attachment of a skeletal muscle to a structure that remains stationary

_____ 8. **motor neuron**  h. a condition that occurs when muscles use more ATP than can be supplied by aerobic respiration

_____ 9. **neurotransmitter**  i. cylindrical structures within a muscle cell that are packed with actin and myosin

_____ 10. **neuromuscular junction**  j. the protein found in thick filaments

_____ 11. **sarcoplasmic reticulum**  k. point of attachment of a skeletal muscle to a structure that moves

_____ 12. **sliding filament mechanism**  l. the junction between a motor neuron and a skeletal muscle cell

_____ 13. **oxygen debt**  m. the protein found in thin filaments

_____ 14. **fatigue**  n. releases acetylcholine

## Word Choice

*Circle the correct term to complete each sentence.*

15. When muscle cells contract they (shorten/lengthen), and when they relax they (shorten/lengthen).

16. Synergistic muscles work to perform (the same/opposing) movements.

17. When a muscle contracts, the (origin/insertion) is pulled toward the (origin/insertion).

18. Using the midline of the body as a reference, the origin of a muscle is usually (closer to/farther from) the midline than the insertion.

19. Maintaining proper posture requires many muscles to (produce/resist) movement.

20. When muscle cells respond to chemical signals they are referred to as (electrical/excitable).

21. A whole muscle is a group of individual muscle cells with (the same/different) origin(s) and insertion(s).

22. The layers of fascia within a single muscle come together at the end of the muscle to form a (tendon/ligament).

23. A single skeletal muscle cell contains (one nucleus/multiple nuclei).

24. A single muscle cell contains (myofibrils packed with actin and myosin/actin and myosin packed with myofibrils).

## Labeling

Label each structure indicated in Figure 6.1.

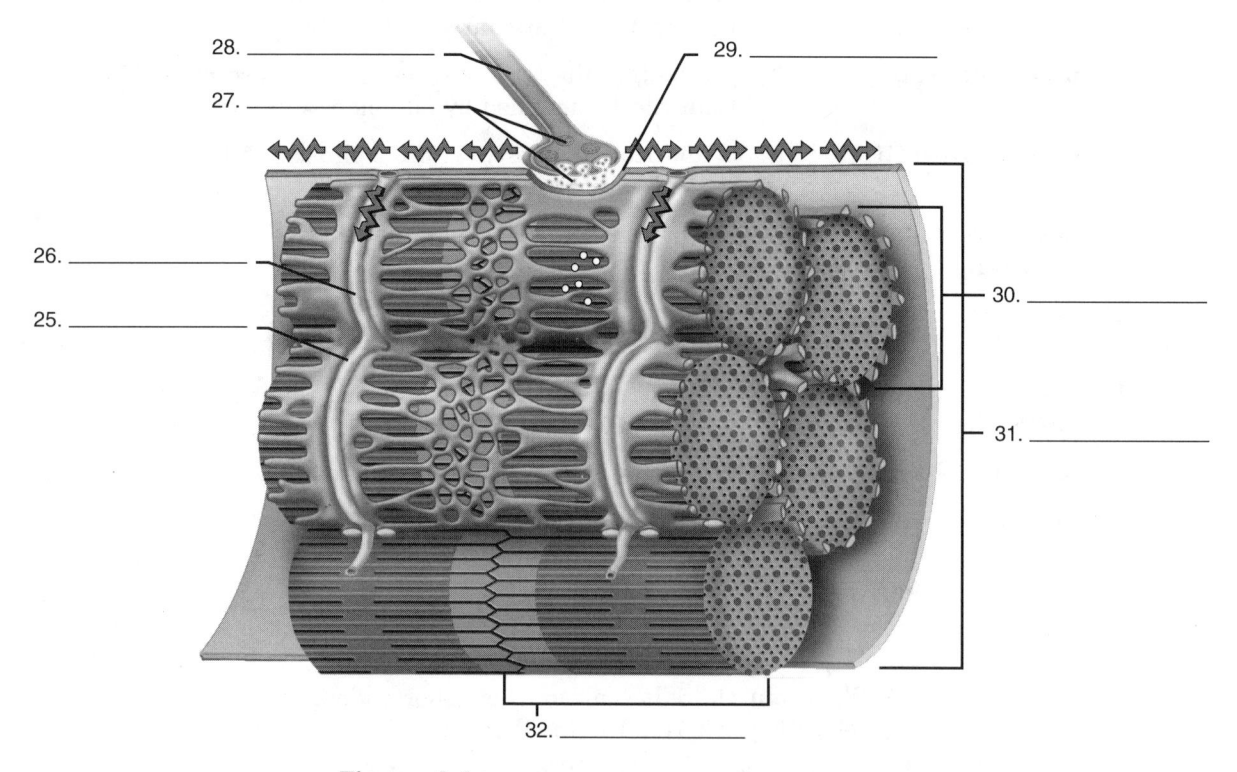

28. _____
27. _____
29. _____
26. _____
25. _____
30. _____
31. _____
32. _____

**Figure 6.1**

## Paragraph Completion

Use the following terms to complete the paragraph, and then repeat this exercise with the terms covered. Terms used more than once are listed multiple times.

electrical
actin
troponin
T tubules
bend
ATP
sliding-filament
acetylcholine

sarcoplasmic reticulum
contraction
lengthens
relaxation
motor neuron
cross-bridges
binding

ATP
neurotransmitter
cross-bridges
sarcoplasmic reticulum
troponin-tropomyosin complex
shortening

The events leading to muscle contraction begin at the neuromuscular junction with the release of (33) _____ from a (34) _____. Acetylcholine is a(n) (35) _____ that attaches to (36) _____ sites on the muscle cell. This causes the muscle cell to generate a(n) (37) _____ impulse that is carried deep into the cell by (38) _____. When the impulse reaches the (39) _____, calcium ions are released into the cytoplasm. The calcium ions bind to the (40) _____ molecule, causing a shift in the position of the (41) _____ and exposing myosin binding sites. With the binding sites exposed, myosin heads form (42) _____ with the actin, then (43) _____ and pull the (44) _____ filament toward the center of the sarcomere. This is called the (45) _____ mechanism of contraction and results in (46) _____ of the sarcomere and (47) _____ of the muscle. (48) _____ is required as an energy source. (49) _____ occurs when nerve stimulation ends. During relaxation, calcium ions are transported back into the (50) _____, and the troponin-tropomyosin complex again blocks the myosin binding sites, preventing the formation of (51) _____. The filaments slide back to their original position and the muscle cell (52) _____. (53) _____ is also required for relaxation.

## Completion

*Answer the following questions about Figure 6.2.*

_____ 54. What represents myosin?

_____ 55. What is associated with the troponin-tropomyosin complex?

_____ 56. What has heads that form cross-bridges?

_____ 57. What slides toward the center of the sarcomere during a contraction?

_____ 58. What has myosin binding sites?

_____ 59. What represents the Z-line?

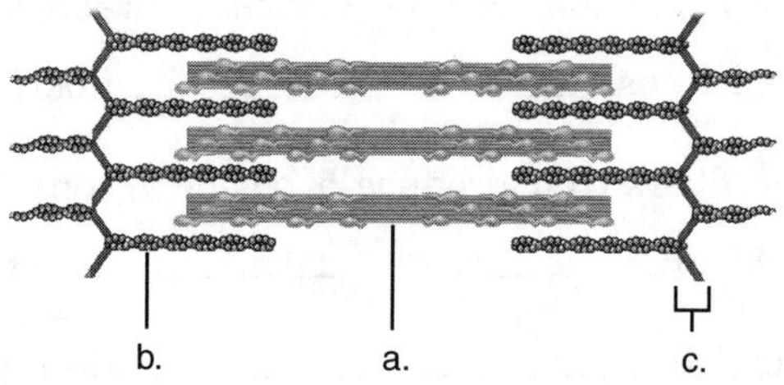

**Figure 6.2**

## Short Answer

60. Explain two ways that ATP is required for muscle relaxation.

61. List the four sources of ATP available to a muscle cell in the order in which the ATP will be accessed.

### 6.3 The activity of muscles can vary

## Matching

____ 1. **motor unit**  a. a complete cycle of contraction and relaxation

____ 2. **muscle tension**  b. increasing muscle force by increasing the rate of stimulation of motor units

____ 3. **all-or-none principle**  c. a motor neuron and all the muscle cells it stimulates

____ 4. **twitch**  d. the mechanical force that muscles generate when they contract

____ 5. **recruitment**  e. muscle cells respond fully every time they are stimulated

____ 6. **summation**  f. increasing muscle force by activating more motor units

## Fill-in-the-Blank

7. Muscle contractions that result in shortening of the muscle are called _____.

8. Muscle contractions that do not result in movement are called _____.

9. Motor units containing only a small number of muscle cells control activities involving _____ muscle control.

10. Muscle _____ is a minimal level of contraction present in most muscles when some motor units are contracting and others are not.

11. Muscle fibers that rely on aerobic metabolism for energy are referred to as _____ fibers.

12. Muscle fibers that produce a powerful, but short, contraction are referred to as _____ fibers.

13. In exercise, _____ training increases the body's efficiency in delivering oxygen to muscles.

## Short Answer

14. Why does summation increase muscle force during a contraction?

15. What causes tetanus?

16. What are the two primary types of exercise training?

17. Why does strength training have a greater effect on fast-twitch fibers than slow-twitch fibers?

18. Why do most marathon runners have a lean build, while weight lifters exhibit greater muscle mass?

19. A friend has just joined the gym and excitedly explains to you that he is taking amino acid supplements to help him make more muscle cells, which he hopes will result in an increase in muscle mass. How would you respond?

### 6.4 Cardiac and smooth muscles have special features

### 6.5 Diseases and disorders of the muscular system

## Matching

*List each muscle type beside all applicable characteristics. Muscle types may be used more than once, and several muscle types may correspond with a characteristic.*

**a. skeletal muscle**     **b. cardiac muscle**     **c. smooth muscle**

_____ 1. voluntary muscle

_____ 2. involuntary muscle

_____ 3. have intercalated discs

_____ 4. striated

_____ 5. non-striated

_____ 6. joined by gap junctions

_____ 7. attached to bones

_____ 8. can only contract when stimulated by a nerve cell

## Short Answer

*For each muscular disorder listed below, indicate the cause and the symptoms.*

9. Muscular Dystrophy:
   a. Cause:
   b. Symptoms:

10. Tetanus:
    a. Cause:
    b. Symptoms:

11. Muscle Cramps:
    a. Cause:
    b. Symptoms:

12. Pulled Muscle:
    a. Cause:
    b. Symptoms:

13. Fasciitis:
    a. Cause:
    b. Symptoms:

---

# Chapter Test

## Multiple Choice

1. Muscle movements over which we have conscious control are:
   a. voluntary.
   b. involuntary.
   c. specific.
   d. natural.

2. The biceps brachii, the brachialis, and brachioradialis all contribute to some of the same movements. These muscles would be:
   a. antagonistic.
   b. synergistic.
   c. complementary.
   d. cooperative.

3. When a skeletal muscle contracts, the _____ is pulled toward the _____.
   a. origin, insertion
   b. origin, muscle
   c. insertion, origin
   d. insertion, muscle

4. Arrange the following components of a muscle in order of increasing size:
   a. fiber, myofibril, fascicle, muscle.
   b. fascicle, myofibril, fiber, muscle.
   c. muscle, myofibril, fascicle, fiber.
   d. myofibril, fiber, fascicle, muscle.

5. When a muscle contracts, shortening of the muscle occurs because:
   a. myofibrils slide over each other.
   b. T tubules shorten.
   c. sarcomeres shorten.
   d. muscle fascia constricts.

6. In the sliding-filament mechanism of muscle contraction, _____ slides over _____.
   a. Z lines, actin
   b. actin, myosin
   c. myosin, actin
   d. Z lines and actin, myosin

7. Which of the following would interfere most with a muscle's ability to contract?
   a. absence of acetylcholine binding sites on the muscle membrane
   b. a reduced number of actin and myosin filaments
   c. an increased level of intracellular calcium
   d. a reduced number of T tubules

8. In a muscle cell _____ contain _____.
   a. myofibrils, actin and myosin filaments
   b. actin and myosin filaments, myofibrils
   c. sarcomeres, myofibrils
   d. myofibrils, sarcomeres
   e. a and d

9. Which of the following best describes the action of sarcomeres within a muscle cell during a contraction?
   a. Each sarcomere shortens a little.
   b. Only the sarcomeres on the ends of the muscle shorten in a weak contraction.
   c. Sarcomeres in the center of the muscle shorten more than sarcomeres on the ends of the muscle.
   d. Sarcomeres shorten in different degrees depending on the amount of actin and myosin they contain.

10. The force of a muscle contraction can be increased by:
    a. increasing the frequency of nerve stimulation.
    b. decreasing the number of motor units responding.
    c. increasing the amount of collagen in the tendons.
    d. decreasing the amount of $Ca^{++}$ released from the sarcoplasmic reticulum.

11. Place the following events of muscle contraction in the proper sequence.
    I. calcium is released from the sarcoplasmic reticulum
    II. cross-bridges form
    III. actin slides across myosin
    IV. calcium binds to troponin, causing the troponin-tropomyosin complex to shift its position
    a. I, II, IV, III
    b. I, IV, II, III
    c. II, I, IV, III
    d. III, I, IV, II

12. In the formation of a cross-bridge, _____ binds to _____.
    a. myosin, actin
    b. actin, troponin
    c. myosin, tropomyosin
    d. tropomyosin, actin

13. When a muscle relaxes, ATP is required. The reason for this is:
    a. the actin filaments must be pushed back to their original position.
    b. the myosin heads can only detach from the actin when a molecule of ATP binds to the head.
    c. energy is needed to pull the sarcomeres apart.
    d. the connective tissue elements of the muscle require energy to resume their original position.

14. Which of the following represents the correct sequence in which a muscle cell will access energy sources?
    a. creatine phosphate, stored ATP, glycogen, aerobic metabolism
    b. aerobic metabolism, creatine phosphate, glycogen, stored ATP
    c. stored ATP, glycogen, creatine phosphate, aerobic metabolism
    d. stored ATP, creatine phosphate, glycogen, aerobic metabolism

15. A muscle contraction that does not result in movement is called:
    a. isometric.
    b. isotonic.
    c. tetanic.
    d. metabolic.

16. The physiological cause of muscle fatigue is:
    a. inadequate sarcomeres to meet muscle demands.
    b. inadequate ATP to meet metabolic demands.
    c. inadequate numbers of slow-twitch fibers.
    d. constriction of blood vessels as lactic acid accumulates.

17. A motor unit consists of:
    a. one motor neuron and all the muscle cells it stimulates.
    b. one muscle and all the motor neurons that stimulate the muscle.
    c. one motor neuron and one muscle cell.
    d. all the motor neurons and all the muscle cells of one whole muscle.

18. The all-or-none principle states that:
    a. when a muscle is stimulated, every motor unit will respond.
    b. when a muscle cell is stimulated, it responds with a complete cycle of contraction and relaxation.
    c. when a muscle cell is stimulated, the sarcomeres that respond will shorten.
    d. a small stimulus can produce a small contraction.

19. Increasing muscle force by increasing the number of motor units responding is called:
    a. summation.
    b. recruitment.
    c. tetany.
    d. magnification.

20. Tetanic contractions result from:
    a. more than one twitch occurring at the same time.
    b. an action potential increasing the size of a muscle's motor units.
    c. myosin and actin being unable to bind to each other.
    d. frequent stimulation preventing a muscle from relaxing.

21. Which of the following is the same in both slow-twitch and fast-twitch muscle fibers?
    a. the all-or-none principle
    b. the role of aerobic metabolism
    c. the number of mitochondria
    d. the arrangement of actin and myosin in a sarcomere
    e. a and d

22. Strength training involves:
    a. improving muscle endurance.
    b. providing resistance that requires muscles to work harder.
    c. increasing the body's ability to supply oxygen to the muscles.
    d. increasing the number of muscle cells.

23. Cardiac muscle and smooth muscle are similar in all of the following ways *except:*
    a. both have muscle cells joined by gap junctions.
    b. the contraction of both can be modified by nerve stimulation.
    c. both are involuntary muscles.
    d. both fatigue easily.

24. Which of the following disorders of the muscular system is incorrectly described?
    a. tetanus—a bacterial disease that causes forceful muscle contractions
    b. muscular dystrophy—a viral disease that causes muscle wasting
    c. pulled muscle—muscle fibers tear apart due to excessive stretching of the muscle
    d. fascitis—inflammation of the connective tissue sheath around a muscle

25. Internal bleeding, swelling, and pain occur with:
    a. tetanus.
    b. muscle cramps.
    c. pulled muscles.
    d. fasciitis.

# Key Concept Review Questions

*Each of the Key Concepts listed at the beginning of this chapter has been rewritten as a question below. After successfully completing the study guide exercises and the Chapter Test, you should be able to answer each of these questions. Refer to the Key Concepts list at the beginning of this chapter to check your answers.*

1. What are three things muscles can do?
2. What are the three types of muscle tissue?
3. How do muscle cells respond when they are stimulated?
4. What structures can be attachment points for muscles?
5. What are the origin and insertion? How do their positions change when a muscle contracts?
6. How would you describe the anatomy of a skeletal muscle?
7. Where are myofibrils located? What do myofibrils contain?
8. What is the contractile unit of a muscle cell?
9. What is the role of a motor neuron in stimulating skeletal muscle cells?
10. What is the effect of acetylcholine on a muscle cell?
11. What causes a sarcomere to shorten? What happens to a muscle cell when the sarcomere shortens?
12. Is ATP required for muscle contraction and relaxation?
13. What affects the force of a muscle contraction?
14. What is a twitch? What precedes a twitch?
15. What is recruitment?
16. What is summation?
17. What characterizes a muscle fiber as slow-twitch or fast-twitch? What qualities of the muscle are affected by the number of slow-twitch and fast-twitch fibers?
18. What are the characteristics of cardiac muscle?
19. What are the characteristics of smooth muscle?
20. What are five disorders of the muscular system described in this chapter?

# Answer Key

## Sections 6.1, 6.2

**1.** d; **2.** g; **3.** k; **4.** i; **5.** a; **6.** j; **7.** m; **8.** n; **9.** e; **10.** l; **11.** f; **12.** b; **13.** h; **14.** c; **15.** shorten, lengthen; **16.** same; **17.** insertion, origin; **18.** closer to; **19.** resist; **20.** excitable; **21.** same; **22.** tendon; **23.** multiple nuclei; **24.** myofibrils packed with actin and myosin; **25.** sarcoplasmic reticulum; **26.** T tubule; **27.** acetylcholine; **28.** motor neuron; **29.** neuromuscular junction; **30.** myofibril; **31.** muscle cell; **32.** sarcomere; **33.** acetylcholine; **34.** motor neuron; **35.** neurotransmitter; **36.** binding; **37.** electrical; **38.** T tubules; **39.** sarcoplasmic reticulum; **40.** troponin; **41.** troponin-tropomyosin complex; **42.** cross-bridges; **43.** bend; **44.** actin; **45.** sliding-filament; **46.** shortening; **47.** contraction; **48.** ATP; **49.** Relaxation; **50.** sarcoplasmic reticulum; **51.** cross-bridges; **52.** lengthens;

**53.** ATP; **54.** a; **55.** b; **56.** a; **57.** b; **58.** b; **59.** c; **60.** Transports Ca$^{++}$ back into the sarcoplasmic reticulum, and allows myosin to detach from actin.; **61.** stored ATP, creatine phosphate, stored glycogen, aerobic metabolism

## Section 6.3

**1.** c; **2.** d; **3.** e; **4.** a; **5.** f; **6.** b; **7.** isotonic; **8.** isometric; **9.** fine; **10.** tone; **11.** slow-twitch; **12.** fast-twitch; **13.** aerobic; **14.** It increases the amount of calcium present; **15.** Stimulating a muscle so frequently that it has no chance to relax; **16.** strength and endurance; **17.** Strength training increases the strength of a particular muscle using resistance; this requires a quick burst of energy, so it targets fast-twitch fibers.; **18.** Running is an aerobic activity requiring the contribution of slow-twitch fibers; slow-twitch fibers do not increase significantly in mass.; **19.** Muscle cells cannot increase in number, however fast-twitch fibers can increase in muscle mass by increasing the number of myofibrils. Strength training, such as weight lifting, can increase muscle mass.

## Sections 6.4, 6.5

**1.** b; **2.** a; **3.** b; **4.** a; **5.** c; **6.** b,c; **7.** a; **8.** a; **9.** a. inherited genetic disorder resulting in the absence of a membrane protein, b. loss of muscle fibers and muscle wasting; **10.** a. bacterial infection, b. tetanic contractions especially affecting the jaw and neck; **11.** a. ATP depletion, dehydration, ion imbalance, lactic acid accumulation, b. uncontrolled muscle contractions, pain; **12.** a. excessive muscle stretching, resulting in torn fibers, b. internal bleeding swelling, pain; **13.** a. straining or tearing of the fascia around a muscle resulting in inflammation, b. pain

## Chapter Test

**1.** a; **2.** b; **3.** c; **4.** d; **5.** c; **6.** b; **7.** a; **8.** e; **9.** a; **10.** a; **11.** b; **12.** a; **13.** b; **14.** d; **15.** a; **16.** b; **17.** a; **18.** b; **19.** b; **20.** d; **21.** e; **22.** b; **23.** d; **24.** b; **25.** c

# 7

# Blood

## Chapter Summary and Key Concepts

*After reading and studying this chapter, you should know the following:*

**Section 7.1**

1. The functions of blood include transport, regulation, and defense.

2. Blood is composed of plasma and formed elements.

3. Plasma is mostly water and also contains ions, proteins, hormones, gases, nutrients, and wastes.

4. Important plasma proteins include albumins that maintain water balance, globulins that transport substances and function as antibodies, and clotting proteins.

5. Formed elements include red blood cells called erythrocytes, white blood cells called leukocytes, and platelets.

6. All formed elements are produced by stem cells in the bone marrow.

7. Mature red blood cells have no nucleus, and they contain an oxygen-binding protein called hemoglobin that transports oxygen from the lungs to the body, and some carbon dioxide from the body to the lungs.

8. The liver and spleen destroy old and damaged red blood cells, recycling the iron atoms and converting heme groups to bilirubin.

9. Red blood cell production is controlled by the hormone erythropoietin.

10. White blood cells defend the body. They are classified as granular or agranular depending on their staining properties.

11. Granular leukocytes include neutrophils, eosinophils, and basophils. Agranular leukocytes include monocytes and lymphocytes.

12. Platelets are essential to the process of blood clotting. Their production is regulated by the hormone thrombopoietin.

### Section 7.2

13. Hemostasis is the process of stopping blood loss, which occurs through vascular spasm, formation of a platelet plug, and coagulation.

14. Formation of a blood clot requires clotting factors, including prothrombin activator, thrombin, and fibrinogen.

### Section 7.3

15. Blood type determines blood compatibility for transfusions.

16. Blood type is determined by cell surface proteins called antigens on the red blood cells.

17. Two common blood typing systems are the ABO system and the Rh system.

18. Transfusion reactions occur when foreign antigens are attacked by antibodies.

### Section 7.4

19. Blood disorders may involve problems with any of the components of the blood.

20. Important blood disorders include anemia, leukemia, multiple myeloma, mononucleosis, blood poisoning, and thrombocytopenia.

---

# Exercises

*Complete the exercises for each section after you have read and studied the section. If you cannot answer some questions, or answer them incorrectly, return to the chapter and review this information. You may find it helpful to work on only one section at a time. When you have completed all sections, take the Chapter Test as an indicator of your mastery of this topic.*

### 7.1   The components and functions of blood

## Matching

\_\_\_\_ 1. **blood**                a.  a cell is consumed and destroyed by a macrophage

\_\_\_\_ 2. **plasma**               b.  red blood cells

\_\_\_\_ 3. **plasma proteins**      c.  hemoglobin with four oxygen molecules attached

\_\_\_\_ 4. **albumins**             d.  type of white blood cell that secretes histamine

\_\_\_\_ 5. **globulins**            e.  cells in the red bone marrow that divide to produce all blood cells and platelets

\_\_\_\_ 6. **erythrocytes**         f.  the most prevalent type of white blood cell

|     |                    |     |                                                                                  |
| --- | ------------------ | --- | -------------------------------------------------------------------------------- |
| ___ | 7. **hemoglobin**       | g.  | the liquid portion of the blood                                                  |
| ___ | 8. **oxyhemoglobin**    | h.  | plasma proteins that transport substances in the blood                           |
| ___ | 9. **deoxyhemoglobin**  | i.  | hemoglobin that has given up its oxygen                                          |
| ___ | 10. **hematocrit**      | j.  | a type of white blood cell that defends the body against large parasites         |
| ___ | 11. **stem cells**      | k.  | large white blood cells that engulf and destroy debris and invaders by phagocytosis |
| ___ | 12. **macrophages**     | l.  | cell fragments involved in blood clotting                                        |
| ___ | 13. **phagocytosis**    | m.  | the largest group of solutes in the plasma                                       |
| ___ | 14. **erythropoietin**  | n.  | a type of white blood cell that develops into a macrophage                       |
| ___ | 15. **blood doping**    | o.  | a type of white blood cell that develops into B cells or T cells                 |
| ___ | 16. **leukocytes**      | p.  | plasma proteins that help maintain water balance in the body                     |
| ___ | 17. **neutrophils**     | q.  | hormone that stimulates the production of red blood cells                        |
| ___ | 18. **eosinophils**     | r.  | artificially increasing the production of red blood cells                        |
| ___ | 19. **basophils**       | s.  | a specialized connective tissue consisting of cells and cell fragments suspended in a watery solution |
| ___ | 20. **monocytes**       | t.  | an oxygen-binding protein found in red blood cells                               |
| ___ | 21. **lymphocytes**     | u.  | white blood cells                                                                |
| ___ | 22. **platelets**       | v.  | the percentage of the blood that consists of red blood cells                     |

## Fill-in-the-Blank

*Referenced sections are indicated in parentheses.*

23. Formed elements of the blood include _____ _____ _____, _____ _____, and _____.(7.1)

24. _____% of whole blood is plasma, and _____% of plasma is water. (7.1)

25. Mature red blood cells lack a _____ and most _____. (7.1)

26. Blood is thicker than water due primarily to the presence of _____ _____ _____. (7.1)

27. Hemoglobin that has given up its oxygen is called _____. (7.1)

28. Worn-out red blood cells are destroyed by the _____ and the _____. (7.1)

29. The percentage of the blood consisting of red blood cells is called the _____. (7.1)

30. Immature cells that give rise to erythrocytes are called _____. (7.1)

31. Platelet production is regulated by the hormone _____. (7.1)

32. The category of agranulocytes that defend the body against large parasites is _____. (7.1)

33. The inflammatory response is initiated by release of _____ from _____. (7.1)

34. Large white blood cells that differentiate into macrophages are _____. (7.1)

35. _____ are agranulocytes important to the immune system and are classified as B cells or T cells. (7.1)

36. When red blood cells are destroyed in the liver, the heme groups are converted to _____. (7.1)

37. Platelets are derived from large cells called _____. (7.1)

## Word Choice

*Circle the accurate term or phrase to complete each sentence.*

38. Mature red blood cells have (no/multiple) nuclei.
39. The bond hemoglobin forms with oxygen is (temporary/permanent).
40. Hemoglobin binds O₂ best when the pH is (neutral/acidic/base).
41. Hemoglobin releases O₂ into the body tissues when the pH of the body tissues are (lower than/higher than) the pH of the blood.
42. Carbon dioxide (attaches to/detaches from) hemoglobin in the lungs.
43. CO₂ and O₂ bind to hemoglobin at (the same site/different sites).
44. Venous blood containing deoxyhemoglobin is (bright red/dark red).

45. Most white blood cells have a (shorter/longer) life span than a red blood cell.
46. Platelet production is regulated by the hormone (thrombopoietin/erythropoietin).
47. (White blood cells/Red blood cells) are part of the body's defense system.

## Completion

*Fill in the requested information for each type of white blood cell listed in the table.*

| Mature White Blood Cell | What type of stem cell gives rise to this cell? | Is this cell granular or agranular? | What function will this cell serve? |
|---|---|---|---|
| Neutrophil | 48. | 49. | 50. |
| Eosinophil | 51. | 52. | 53. |
| Basophil | 54. | 55. | 56. |
| Monocyte | 57. | 58. | 59. |
| Lymphocyte | 60. | 61. | 62. |

## Ordering

*Place these events that occur during the negative feedback control of red blood cell production in the correct sequence, with "1" being the first event and "7" being the last event.*

_____ 63. erythropoietin stimulates stem cells to produce more red blood cells

_____ 64. blood oxygen levels fall

_____ 65. kidney cells decrease production of erythropoietin

_____ 66. blood oxygen levels rise

_____ 67. cells in the kidney constantly monitor the availability of oxygen in the blood

_____ 68. erythropoietin is transported to the red bone marrow

_____ 69. kidney cells secrete erythropoietin into the bloodstream

## Short Answer

70. Explain how structure and function are related in red blood cells.

71. Describe the conditions under which hemoglobin binds oxygen most efficiently. Where do these conditions exist?

72. Compare and contrast red blood cells and white blood cells by listing two ways that they are similar and two ways that they are different.

### 7.2 Hemostasis: Stopping blood loss

## Paragraph Completion

*Use the following terms to complete the paragraph, and then repeat this exercise with the terms covered.*

| prothrombin | hemostasis | platelet plug |
| coagulation | prothrombin activator | fibrin |
| hemophilia | vessel | clot |
| vascular spasm | clotting factors | thrombin |
| fibrinogen | contract | platelets |

The natural process of stopping the loss of blood from ruptured vessels is called (1)_____ and occurs in three stages. The first stage is (2)_____, during which blood vessels (3)_____ to reduce blood flow. The second stage involves formation of a (4)_____, in which platelets stick to other (5)_____ and to (6)_____ walls, sealing the injured area. The third stage is (7)_____. Substances known as (8)_____ participate in the formation of a blood clot. Damage to vessels stimulates the release of (9)_____, which converts (10)_____ to the enzyme (11)_____. This enzyme in turn converts the plasma protein (12)_____ into long, soluble protein threads called (13)_____. These threads interact with platelets and trapped red blood cells to form a (14)_____. Individuals with (15)_____, an inherited condition, lack one or more clotting factors and are unable to form effective blood clots.

### 7.3 Human blood types

## Fill-in-the-Blank

*Referenced sections are indicated in parentheses.*

1. The success of a blood transfusion depends largely on _____ _____. (7.3)

2. Cells have surface _____ that are recognized by the immune system as "self." (7.3)

3. Foreign cells carry surface proteins called _____ that are recognized by the immune system as "nonself." (7.3)

4. When the immune system is exposed to foreign cells it will mount a defense that includes the production of _____ by _____. (7.3)

5. Antibodies float freely in the _____ and _____. (7.3)

6. When antibodies encounter a foreign antigen they will bind to it, forming a(n) _____. (7.3)

7. Throughout life, individuals carry circulating _____ against any antigens different from their own. (7.3)

8. When red blood cells with foreign antigens are attacked by _____, they clump together or _____. (7.3)

9. Clumping together of red blood cells (RBCs) may lead to _____ _____ or _____. (7.3)

10. An adverse effect of a blood transfusion is called a _____. (7.3)

## Short Answer

11. An individual has Type B blood. What type of antigen is found on their red blood cells? What type of antibodies do they carry?

12. An individual has Type O blood. What type of antigen is found on their red blood cells? What type of antibodies do they carry?

13. Two individuals both have blood type A; however, one is type A+ and the other is type A−. What is the difference in their red blood cells?

14. Why can an individual with Rh− blood receive a transfusion of Rh+ blood one time without complications, while a second transfusion of Rh+ blood may cause a severe reaction?

15. a. What is hemolytic disease of the newborn? b. What parental blood types would put a newborn at risk for this disease? c. Why does this disease seldom affect a first child? d. How can it be prevented in subsequent children?

16. Describe the process of cross-matching. What results indicate that two blood samples are a good match?

### 7.4 Blood disorders

## Matching

____ 1. **mononucleosis**    a. infection of lymphocytes caused by the Epstein-Barr virus

____ 2. **anemia**    b. uncontrolled production of abnormal or immature white blood cells

____ 3. **multiple myeloma**    c. a reduction of the number of platelets in the blood

____ 4. **leukemia**    d. a reduced ability of the blood to transport oxygen due to a variety of causes

____ 5. **blood poisoning**    e. uncontrolled division of abnormal plasma cells in the bone marrow

____ 6. **thrombocytopenia**    f. proliferation of microorganisms in the blood plasma

# Chapter Test

## Multiple Choice

1. The circulatory system consists of:
    a. red blood cells, white blood cells, and platelets.
    b. plasma and formed elements.
    c. blood vessels and lymphatic vessels.
    d. heart, blood vessels, and blood.

2. Which of the following is not a function of the blood?
    a. protection against infection and illness
    b. regulation of body temperature and fluid balance
    c. secretion of erythropoietin
    d. transportation

3. Which of the following is *not* a component of plasma?
    a. platelets
    b. water
    c. dissolved protein
    d. ions

4. Which of the following plasma proteins is incorrectly matched with its function?
    a. globulins—transport substances in the blood
    b. clotting proteins—maintain fluidity of the blood by preventing clotting
    c. albumins—maintain the osmotic relationship between blood and the interstitial fluid
    d. a, b, and c are correctly matched

5. LDLs are:
    a. a beneficial type of cholesterol.
    b. a type of clotting protein.
    c. a lipoprotein associated with heart disease.
    d. cell surface antigens.

6. The iron atom of a heme group functions to:
    a. bind the heme group to the polypeptide chains of hemoglobin.
    b. bind oxygen.
    c. keeps red blood cells suspended in plasma.
    d. extend the life of a mature red blood cell in the absence of a nucleus.

7. Which of the following conditions increases the release of oxygen from hemoglobin?
    a. high concentration of dissolved oxygen in the body tissues
    b. increased pH in the tissues
    c. increased body heat
    d. decreased metabolic rate

8. Red blood cells are accurately described as:
    a. large and oblong cells.
    b. rigid and inflexible cells.
    c. small and flat cells.
    d. multinucleate cells.

9. Stem cells in the bone marrow produce:
    a. red blood cells only.
    b. white blood cells only.
    c. red and white blood cells.
    d. red cells, white cells, and platelets.

10. Red blood cells have a short life span because:
    a. they are easily damaged as they pass through the blood vessels.
    b. they are often destroyed by immune system cells.
    c. they lack a nucleus.
    d. their cell membranes are impermeable to required nutrients.

11. Your friend shows you a yellow bruise on her arm. You explain to her that the yellow color is due to the:
    a. reduced oxygen-carrying capacity of damaged red blood cells.
    b. breakdown of heme as red blood cells are destroyed.
    c. increased oxygen binding capacity of hemoglobin during healing.
    d. high concentration of carbon dioxide at the site of the bruise.

12. Erythropoietin stimulates production of _____ in response to _____.
    a. white blood cells, infection
    b. red blood cells, low $O_2$ availability
    c. platelets, blood vessel damage
    d. fibrin, increased fibrinogen levels

13. Basophils:
    a. defend the body against large parasites.
    b. release histamine when tissues are injured.
    c. are classified as B cells or T cells.
    d. are the first white blood cells to combat infection.

14. All of the following are granular leukocytes *except:*
    a. neutrophils.
    b. monocytes.
    c. eosinophils.
    d. basophils.

15. Which of the following conditions would increase the release of $O_2$ from red blood cells into the tissue fluids?
    a. neutral pH
    b. high concentration of $O_2$ in the tissue fluids
    c. low concentration of $O_2$ in the tissue fluids
    d. low concentration of $CO_2$ in the tissue fluids

16. In the process of hemostasis, when a blood vessel breaks:
    a. red blood cells differentiate into macrophages.
    b. eosinophils migrate to the site to seal the rupture.
    c. platelets swell, develop spiky extensions, and begin to clump together.
    d. the kidneys decrease production of erythropoietin.

17. In what order do the following substances participate in the formation of a blood clot?
    I. fibrinogen
    II. thrombin
    III. prothrombin activator
    IV. fibrin
    V. prothrombin
    a. IV, I, V, III, II
    b. I, IV, III, V, II
    c. II, V, III, IV, I
    d. III, V, II, I, IV

18. An individual who has blood type AB+ will have _____ antigens on the surface of their red blood cells.
    a. A and B
    b. only Rh
    c. A, B, and Rh
    d. no

19. Agglutination between red blood cells of two blood samples indicates:
    a. the two samples are compatible.
    b. the cells of both samples have the same antigens.
    c. the two samples are not compatible.
    d. the antigens of one sample have attacked the antibodies of the other sample.

20. During a transfusion reaction:
    a. recipient antigens attack donor antibodies.
    b. recipient antigens fail to bind to donor antigens.
    c. recipient antibodies attack donor antigens.
    d. recipient antibodies engulf and destroy foreign red blood cells by phagocytosis.

21. In the ABO blood system, antibodies appear in an individual's blood:
    a. only in response to foreign antigens.
    b. early in life, with or without exposure to foreign antigens.
    c. only during childhood.
    d. only in response to Rh+ blood.

22. Which of the following is not true of leukemia?
    a. it is a type of blood cancer
    b. too many white blood cells are produced
    c. the rate of red blood cell destruction is increased
    d. organ function is affected

23. Which type of anemia results from a deficiency in vitamin $B_{12}$ absorption by the digestive tract?
    a. iron deficiency anemia
    b. sickle cell anemia
    c. aplastic anemia
    d. pernicious anemia

24. Mononucleosis:
    a. is not contagious.
    b. is the result of a bacterial infection.
    c. has no known cure.
    d. symptoms include easy bruising and bone pain.

25. An individual who has blood type A can receive a transfusion of which blood types?
    a. only type A
    b. type A and type O
    c. type A and type B
    d. type A and type AB
    e. type A, type AB, and type O

## Key Concept Review Questions

*Each of the Key Concepts listed at the beginning of this chapter has been rewritten as a question below. After successively completing the study guide exercises and the Chapter Test, you should be able to answer each of these questions. Refer to the Key Concepts list on the first page of this chapter to check your answers.*

1. What are the functions of the blood?
2. What are the two main components of blood?
3. What is plasma composed of?
4. What are three important categories of plasma proteins and their functions?
5. What are the formed elements of the blood?
6. Where are formed elements produced? What type of cell is involved?
7. What is found inside red blood cells? What is the function of hemoglobin?
8. Where and how are old red blood cells destroyed?
9. What hormone controls the production of red blood cells?
10. What is the function of white blood cells? How are they classified?
11. Which white blood cells are classified as granular leukocytes? Which white blood cells are classified as agranular leukocytes?
12. What process requires platelets? What hormone regulates platelet production?
13. What is hemostasis? What are the stages of hemostasis?
14. What substances are required for the formation of a blood clot?
15. What determines blood compatibility?
16. What determines an individual's blood type?
17. What are two commonly used blood typing systems?
18. What causes a transfusion reaction?
19. What parts of the blood may be involved in blood disorders?
20. What are seven blood disorders discussed in Chapter 7?

## Answer Key

### Section 7.1

**1.**s; **2.**g; **3.**m; **4.**p; **5.**h; **6.**b; **7.**t; **8.**c; **9.**i; **10.**v; **11.**e; **12.**k; **13.**a; **14.**q; **15.**r; **16.**u; **17.**f; **18.**j; **19.**d; **20.**n; **21.**o; **22.**l; **23.** red blood cells, white blood cells, platelets; **24.** 55, 90; **25.** nucleus, organelles; **26.** red blood cells; **27.** deoxyhemoglobin; **28.** liver, spleen; **29.** hematocrit; **30.** erythroblasts;

31. thrombopoietin; 32. eosinophils; 33. histamine, basophils; 34. monocytes; 35. Lymphocytes; 36. bilirubin; 37. megakaryocytes; 38. no; 39. temporary; 40. neutral; 41. lower than; 42. detaches from; 43. different sites; 44. dark red; 45. shorter; 46. thrombopoietin; 47. White blood cells; 48. myeloblast; 49. granular; 50. first cells to combat infection, they engulf and destroy foreign cells; 51. myeloblast; 52. granular; 53. defend the body against large parasites and release chemicals that affect allergic reactions; 54. myeloblast; 55. granular; 56. secrete histamine; 57. monoblast; 58. agranular; 59. differentiate into macrophages and stimulate lymphocytes; 60. lymphoblast; 61. agranular; 62. critical to the body's immune system; 63. 5; 64. 2; 65. 7; 66. 6; 67. 1; 68. 4; 69. 3; 70. the flat rounded red blood cells are flexible enough to squeeze through the smallest blood vessels, and their flat structure provides for efficient gas exchange; 71. high concentration of oxygen and neutral pH, lungs; 72. similarities: both are produced in red bone marrow, both are components of the blood; differences: rbcs lack a nucleus while wbcs have a nucleus, rbcs transport oxygen and carbon dioxide while wbcs function in defending the body, rbcs remain in the bloodstream while wbcs migrate out of the blood and through the tissues

## Section 7.2

1. hemostasis; 2. vascular spasms; 3. contract; 4. platelet plug; 5. platelets; 6. vessel; 7. coagulation; 8. clotting factors; 9. prothrombin activator; 10. prothrombin; 11. thrombin; 12. fibrinogen; 13. fibrin; 14. clot; 15. hemophilia

## Section 7.3

1. blood type; 2. proteins; 3. antigens; 4. antibodies, lymphocytes; 5. blood, lymph; 6. antigen-antibody complex; 7. antibodies; 8. antibodies, agglutinate; 9. organ damage, death; 10. transfusion reaction; 11. B, A; 12. neither A nor B, A and B; 13. type A+ also has Rh antigens; 14. antibodies are produced only after exposure; 15. a. Rh antibodies spread from mother to the fetus and attack fetal red blood cells, b. Mother: Rh−, Father: Rh+, c. no maternal antibodies are present yet, d. maternal injection of RhoGAM; 16. samples of donor blood are mixed with recipient plasma and samples of recipient blood are mixed with donor plasma, absence of agglutination in both combinations

## Section 7.4

1.a; 2.d; 3.e; 4.b; 5.f; 6.c

## Chapter Test

1.d; 2.c; 3.a; 4.b; 5.c; 6.b; 7.c; 8.c; 9.d; 10.c; 11.b; 12.b; 13.b; 14.b; 15.c; 16.c; 17.d; 18.c; 19.c; 20.c; 21.b; 22.c; 23.d; 24.c; 25.b

# 8

# Heart and Blood Vessels

## Chapter Summary and Key Concepts

*After reading and studying this chapter you should know the following:*

**Sections 8.1, 8.2**

1. The pumping action of the heart provides the power to transport blood.

2. Arteries transport blood away from the heart; veins transport blood toward the heart.

3. Capillaries transport blood from arteries to veins, and they allow for exchange of fluid, nutrients, and wastes with the tissues.

4. The structure of the vessel wall is suited to its function.

5. Veins return blood to the heart with the aid of skeletal muscle activity, one-way valves, and pressure changes associated with breathing.

6. The lymphatic system helps to maintain blood volume by returning excess extracellular fluid to the blood.

7. The heart consists of four chambers—two upper atria that receive blood, and two lower ventricles that eject blood from the heart.

8. Heart valves prevent blood from flowing backward.

9. The right side of the heart receives and then pumps deoxygenated blood. The left side of the heart receives and then pumps oxygenated blood.

10. The pulmonary circuit transports blood between the heart and the lungs for oxygenation.

11. The systemic circuit transports blood between the heart and the rest of the body to supply oxygen and nutrients to the tissues and remove wastes from the tissues.

12. The cardiac cycle includes periods of contraction called systole, and periods of relaxation called diastole.

13. Electrical impulses cause contraction of heart muscle, and they are initiated and spread through the heart by the cardiac conduction system.

### Sections 8.3, 8.4, 8.5, 8.6

14. Blood pressure is the force that blood exerts on the vessel walls.
15. Blood pressure is an important indicator of the overall health of the cardiovascular system.
16. Systolic pressure is produced when ventricles contract, and diastolic pressure is produced when ventricles relax.
17. A constant arterial blood pressure is crucial to homeostasis, and is maintained by regulating heart rate, force of contraction of the heart, and the diameter of arterioles.
18. Regulation of the cardiovascular system is accomplished by receptors in the arteries, regulatory centers in the brain, hormones, and the environment in the tissues.
19. Cardiovascular disorders include hypertension, angina pectoris, heart attack, congestive heart failure, embolism, and stroke.
20. There are many voluntary actions that can help to maintain cardiovascular health.

# Exercises

Complete the exercises for each section after you have read and studied the section. If you cannot answer some questions, or answer them incorrectly, return to the chapter and review this information. You may find it helpful to work on only one section at a time. When you have completed all sections, take the Chapter Test as an indicator of your mastery of this topic.

**8.1   Blood vessels transport blood**

**8.2   The heart pumps blood through the vessels**

## Matching

\_\_\_\_  1. **cardiovascular system**        a.  band of smooth muscle that controls blood flow into the capillaries

\_\_\_\_  2. **arteries**                     b.  the outermost layer of the heart

\_\_\_\_  3. **arterioles**                   c.  valves that prevent blood from flowing back into the atria when the ventricles contract

\_\_\_\_  4. **precapillary sphincter**       d.  pathway of blood flow that sends deoxygenated blood to the lungs and returns oxygenated blood to the heart

\_\_\_\_  5. **capillaries**                  e.  the inner layer of the heart

\_\_\_\_  6. **veins**                        f.  period of relaxation of cardiac muscle

\_\_\_\_  7. **epicardium**                   g.  the layer of the heart wall consisting mainly of cardiac muscle

\_\_\_\_  8. **myocardium**                   h.  period of contraction in the cardiac muscle

____ 9. **endocardium**  i. record of electrical impulses in the cardiac conduction system

____ 10. **atria**  j. vessels that transport blood from the body back to the heart

____ 11. **ventricles**  k. sinoatrial (SA) node

____ 12. **septum**  l. specialized cardiac muscle cells that initiate and distribute electrical impulses through the heart

____ 13. **atrioventricular valves (AV)**  m. the largest artery in the body

____ 14. **semilunar valves**  n. pathway of blood flow that carries oxygenated blood to the body and returns deoxygenated blood to the heart

____ 15. **pulmonary circuit**  o. the two upper chambers of the heart

____ 16. **systemic circuit**  p. body system consisting of the heart and blood vessels

____ 17. **pulmonary veins**  q. blood vessels that supply the heart muscle

____ 18. **aorta**  r. the smallest arteries of the body

____ 19. **systole**  s. the muscular partition separating the right and left sides of the heart

____ 20. **diastole**  t. the smallest vessels whose walls consists of a single layer of cells

____ 21. **cardiac cycle**  u. vessels that carry oxygenated blood to the heart from the lungs

____ 22. **cardiac conduction system**  v. thick-walled vessels that transport blood at high pressure away from the heart

____ 23. **cardiac pacemaker**  w. valves that prevent blood from flowing backward into the ventricles when the heart relaxes

____ 24. **electrocardiogram (ECG)**  x. the two lower chambers of the heart

____ 25. **coronary arteries**  y. the complete sequence of contraction and relaxation in the atria and ventricles

## Fill-in-the-Blank

*Referenced sections are indicated in parentheses.*

26. The three major types of blood vessels in the body are _____, _____, and _____. (8.1)

27. Water and small solutes escape from a capillary through _____ in the epithelial cells and _____ between the cells. (8.1)

28. Ballooning of the arterial wall due to vessel damage is referred to as a(n) _____. (8.1)

29. Contraction of vascular smooth muscle is called _____. (8.1)

30. _____ of arterioles and precapillary sphincters increases blood flow to the capillaries. (8.1)

31. Fluid filtered from the beginning of a capillary is reabsorbed by _____ into the last part of the capillary. (8.1)

32. _____ _____ are extensive networks of capillaries found in all areas of the body. (8.1)

33. Filtration of fluid out of the capillaries is caused by _____ _____ generated by the heart. (8.1)

34. Blind-ended vessels that pick up excess extracellular fluid are _____ _____. (8.1)

35. The fluid carried by lymphatic vessels is called _____. (8.1)

36. Blood flows from capillaries into _____. (8.1)

37. Veins permanently swollen by pooled blood are called _____ veins. (8.1)

38. The _____ is a tough fibrous sac surrounding the heart. (8.2)

39. The _____ _____ contains fluid that provides a lubricant for contraction of cardiac muscle. (8.2)

40. An abnormality in the heart rate is called a(n) _____. (8.2)

41. Inflammation of the innermost layer of the heart is called _____. (8.2)

42. During an ECG, the end of electrical activities in the ventricles is indicated by the _____. (8.2)

43. The _____ of an ECG immediately precedes ventricular contraction. (8.2)

# Labeling

*Label each indicated structure in Figure 8.1 with all applicable structures and functions listed below the diagram. Structures and functions may be used more than once.*

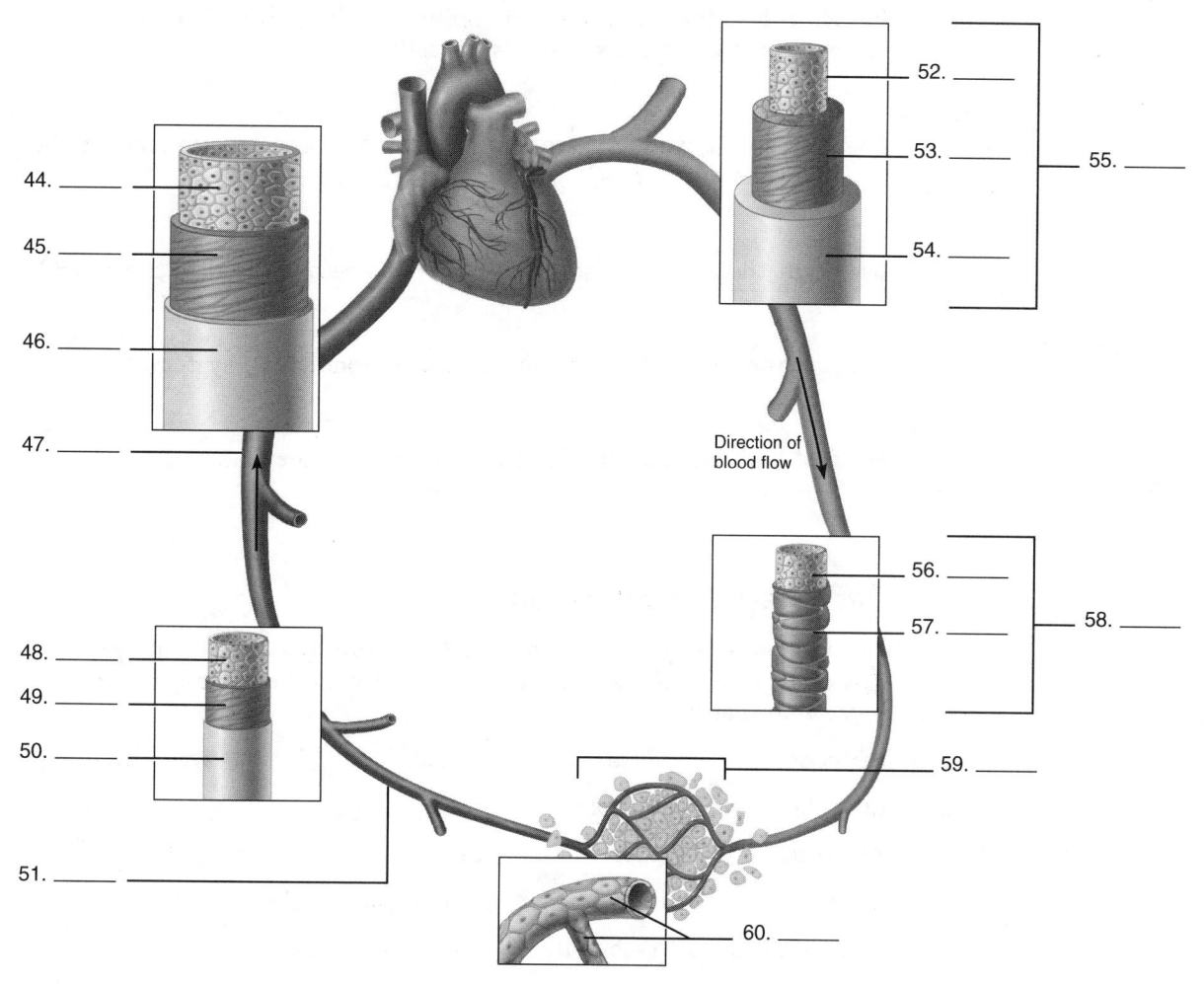

**Figure 8.1**

a. vein
b. capillary
c. arteriole
d. venule
e. artery
f. endothelium
g. capillary bed
h. smooth muscle
i. outer layer of connective tissue
j. red blood cells often pass through these vessels single file
k. continuation of the lining of the heart
l. blood pressure in these vessels is very small
m. this allows these vessels to withstand high pressure
n. the blood vessel wall is a single layer of squamous epithelial cells
o. these vessels contain valves that keep blood flowing in one direction
p. these vessels have a large lumen
q. this allows these vessels to stretch in response to pressure
r. these blood vessels help to regulate the amount of blood that flows to each capillary
s. anchors blood vessels to surrounding tissues
t. the thickest layer in most arteries

## Short Answer

61. List three functions of arterioles.

62. How does the structure of a capillary wall enable blood to exchange substances with tissue cells?

63. Why do the walls of veins not require as much muscle strength as the walls of arteries?

64. What three mechanisms help the veins return blood to the heart?

65. List the four structures of the cardiac conduction system.

66. List and describe the three layers of the heart wall.

## Paragraph Completion

*Use the following terms to complete the paragraph, and then repeat this exercise with the terms covered. Terms used more than once are listed multiple times.*

| SA node | aorta | atrium | atrial systole |
| diastole | diastole | systole | pulse |
| AV node | atria | AV | systole |
| diastole | conduction | ventricular systole | atria |
| heart sounds | AV bundle | pulmonary trunk | |
| ventricles | brain | ventricles | |

A complete cardiac cycle involves contraction of the two (67)_____, which forces blood into the ventricles, followed by contraction of the two (68)_____, which pumps blood into the pulmonary arteries and the aorta. The period of contraction is referred to as (69)_____, and the period of relaxation is referred to as (70)_____. The cardiac cycle begins with (71)_____, when all chambers of the heart are relaxed. Blood flows from the veins into the (72)_____ and then passively into the (73)_____. The next stage is (74)_____ when the atria contract, forcing any remaining blood into the ventricles.

The final stage is (75)_____, which closes the two (76)_____ valves and forces blood into the (77)_____ and (78)_____. Artery walls stretch during (79)_____ and recoil during (80)_____; this can be felt in arteries near the skin surface as a (81)_____. The closing of the valves during the cardiac cycle generates (82)_____. Coordination of the cardiac cycle is due to the cardiac (83)_____ system. Electrical impulses are initiated in the (84)_____ located in the right (85)_____. Impulses travel across the heart, next reaching the (86)_____. In the septum between the two ventricles lie conducting fibers called the (87)_____, which branch and carry the electrical impulses throughout the ventricles. In addition to impulses initiated by the heart's own pacemaker, heart rate can be influenced by messages from the (88)_____.

## Ordering

*Order the following structures to indicate the correct sequence of blood flow, beginning and ending with the right atrium.*

| | | |
|---|---|---|
| left atrioventricular valve | venules | veins |
| right ventricle | right atrioventricular valve | pulmonary trunk |
| aortic semilunar valve | pulmonary semilunar valve | arterioles |
| lungs | capillaries | pulmonary veins |
| aorta | left atrium | arteries |
| | | left ventricle |

89. right atrium →a. _____ →b. _____
→c. _____ →d. _____ →e. _____
→f. _____ →g. _____ →h. _____
→i. _____ →j. _____ →k. _____
→l. _____ →m. _____ →n. _____
→o. _____ →p. _____ →right atrium

## Labeling

90. Label Figure 8.2 with the words listed below.

| | |
|---|---|
| pulmonary trunk | pulmonary semilunar valve |
| aortic semilunar valve | right ventricle |
| left AV valve | right AV valve |
| aorta | septum |
| right atrium | left ventricle |
| left atrium | |

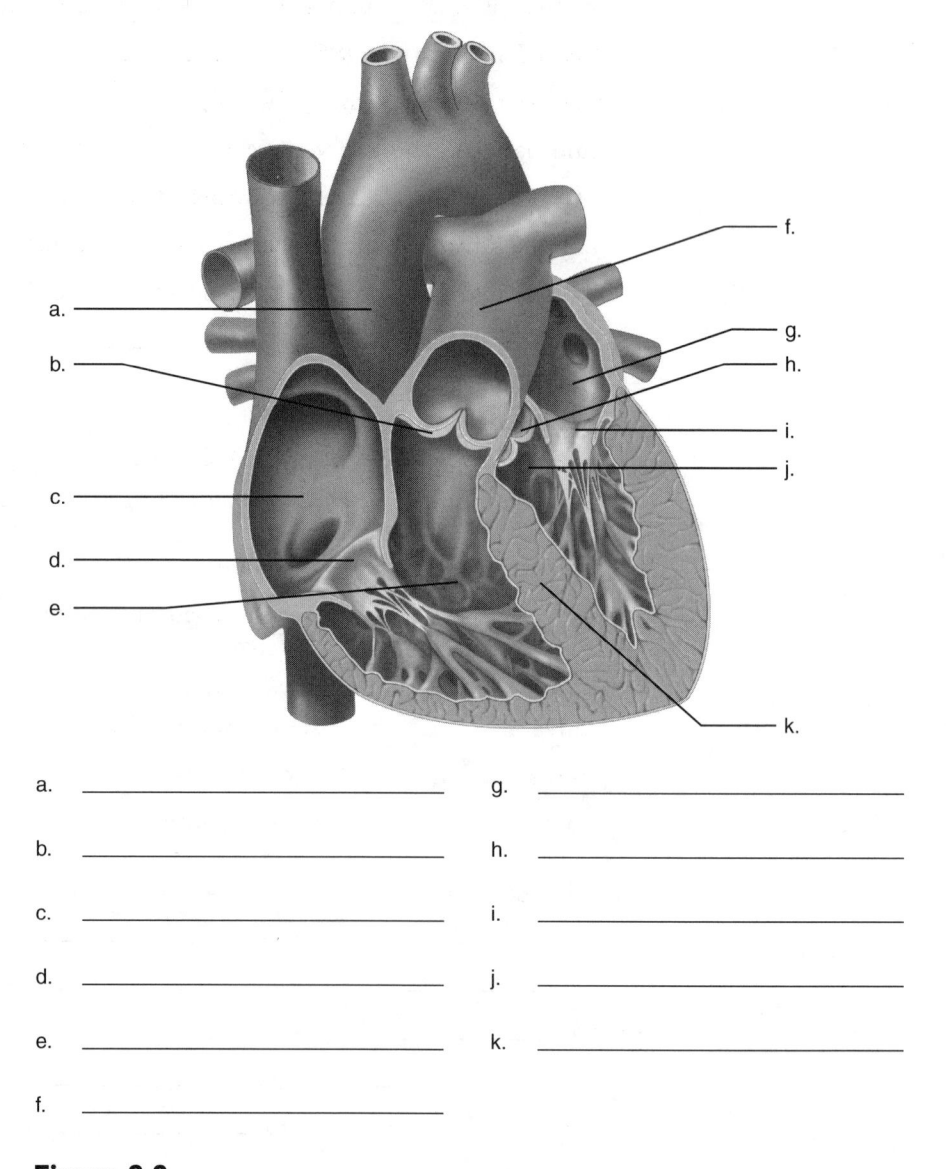

a. _____  g. _____

b. _____  h. _____

c. _____  i. _____

d. _____  j. _____

e. _____  k. _____

f. _____

**Figure 8.2**

8.3 Blood exerts pressure against vessel walls
8.4 How the cardiovascular system is regulated
8.5 Cardiovascular disorders: A major health issue
8.6 Reducing your risk of cardiovascular disease

## Matching

_____ 1. **blood pressure**  a. special structures in the arteries that monitor and help to regulate blood pressure

_____ 2. **systolic pressure**  b. death of heart tissue due to lack of oxygen

_____ 3. **diastolic pressure**  c. the lowest blood pressure, occurs during ventricular relaxation

_____ 4. **hypertension**  d. the amount of blood the heart pumps into the aorta in 1 minute

_____ 5. **baroreceptors**  e. damage to brain tissue caused by interruption of the blood supply

_____ 6. **cardiac output**  f. buildup of extracellular fluid due to weakening of heart muscle

_____ 7. **heart attack**  g. sudden blockage of a blood vessel, often due to a blood clot

_____ 8. **congestive heart failure**  h. high blood pressure

_____ 9. **embolism**  i. force exerted by blood on the walls of the blood vessels

_____ 10. **stroke**  j. highest blood pressure, occurs during ventricular contraction

## Fill-in-the-Blank

*Referenced sections are indicated in parentheses.*

11. Blood pressure is measured with a _____. (8.3)

12. In measuring blood pressure, the pressure at which blood reenters the collapsed artery is the _____ pressure. (8.3)

13. When diastolic pressure remains normal while systolic pressure is too high, this is called _____ _____. (8.3)

14. _____ is a condition when blood pressure falls below normal. (8.3)

15. Cardiac output is determined by multiplying _____ _____ by _____ _____. (8.4)

16. _____ nerves from the _____ _____ of the brain can stimulate the heart to beat faster. (8.4)

17. Cardiovascular disorders are the number _____ killer in the United States. (8.5)

18. A(n) _____ is an x-ray of the blood vessels. (8.5)

19. _____ is a surgical procedure in which blood vessels are taken from one part of the body and grafted onto a blocked coronary artery. (8.5)

## Short Answer

20. Why is blood pressure lower in the arterioles and capillaries than in the arteries?

21. How does blood flow in the capillaries differ from blood flow in the arteries?

22. How do blood pressure differences in the arteries, capillaries, and veins keep blood flowing through the body?

## Paragraph Completion

*Use the following terms to complete the paragraph, and then repeat this exercise with the terms covered.*

| arterial | brain | parasympathetic | carotid arteries |
| increase | heart rate | epinephrine | sympathetic |
| dilate | oxygen | carbon dioxide | negative feedback |
| diameter | aorta | adrenal glands | medulla oblongata |

Homeostatic regulation of the cardiovascular system depends on maintaining a constant (23)_____ blood pressure. This regulation is accomplished by the activity of several structures. Baroreceptors, located in the (24)_____ and (25)_____, stretch as blood pressure rises. This sends signals to the cardiovascular center in the (26)_____. The response is lowering of the (27)_____, which reduces cardiac output. Signals received by the brain in response to falling blood pressure trigger a(n) (28)_____ in cardiac output and constriction of the arterioles. This activity between the baroreceptors and the brain represents a (29)_____ loop. The cardiovascular center is located in the (30)_____ of the brain. Nerves from this area can influence heart activity. (31)_____ nerves stimulate the heart and increase heart rate, while (32)_____ nerves inhibit the heart. These nerves can also affect blood pressure by changing the (33)_____ of the arterioles. The hormone (34)_____ produced by the (35)_____ can stimulate the heart in times of stress. Blood flow in a particular area is regulated by local tissue needs. When tissues are metabolically active, they consume more (36)_____ and produce more (37)_____. This causes precapillary sphincters to (38)_____, increasing blood flow to the area.

## Short Answer

*For each cardiovascular system disorder listed below, indicate the cause and the symptoms.*

39. Hypertension
    a. cause: _____
    b. symptoms: _____

40. Angina pectoris
    a. cause: _____
    b. symptoms: _____

41. Heart attack
    a. cause: _____
    b. symptoms: _____

42. Congestive heart failure
    a. cause: _____
    b. symptoms: _____

43. Embolism
    a. cause: _____
    b. symptoms: _____

44. Stroke
    a. cause: _____
    b. symptoms: _____

# Chapter Test

## Word Choice

*Circle the accurate term to complete each sentence.*

1. Vasodilation of arterioles (increases/decreases) their diameter, (increasing/decreasing) blood flow to the capillaries.
2. The walls of veins are (thinner/thicker) than the walls of arteries.
3. The lumen of a vein is (larger/smaller) in diameter than the lumen of an artery.
4. Blood pressure in the veins is (less than/greater than) blood pressure in the arteries.
5. During diastole, pressure inside the heart chambers is (less than/greater than) blood pressure in the veins.
6. Baroreceptors stretch as blood pressure (increases/decreases).
7. Blood entering the right atrium is (oxygenated/deoxygenated).

## Multiple Choice

8. The heart has _____ chambers and _____ valves.
   a. 2, 2
   b. 2, 4
   c. 4, 2
   d. 4, 4

9. In the functioning of the cardiovascular system, the heart provides _____, while the blood vessels provide _____.
   a. new red blood cells, oxygen
   b. baroreceptors to monitor blood pressure, smooth muscle capable of changing vessel diameter to respond to blood pressure
   c. power to move the blood, a network through which blood can flow
   d. hormones to regulate cardiac output, receptors to monitor cardiac output

10. Large- and medium-sized arteries are:
    a. stiff, yet elastic.
    b. thin, yet strong.
    c. thick, yet diffusible.
    d. none of the above.

11. Which layer of a large blood vessel's wall will be primarily smooth muscle?
    a. inner layer
    b. middle layer
    c. outer layer
    d. epithelial layer

12. Which of the following accurately describes arterioles?
    a. they have the same outer layer of connective tissue as that found in arteries
    b. they deliver blood directly to a venule
    c. they help regulate the amount of blood that flows to each capillary
    d. blood pressure in the cardiovascular system is highest in the arterioles

13. Capillary walls consist of:
    a. a thin layer of smooth muscle surrounding the endothelium.
    b. smooth muscle enclosed by a thin layer of connective tissue.
    c. a single layer of endothelium composed of epithelial cells.
    d. three layers similar to larger arteries, but each layer is much thinner.

14. Excess extracellular fluid accumulates in the tissues because:
    a. the capillary wall is too thick to allow the fluid to diffuse back into the vessels.
    b. the extracellular fluid contains too many proteins to allow return to the capillary.
    c. the pressure-induced filtration of water is greater than the diffusional reabsorption of water.
    d. the extracellular fluid provides storage for excess oxygen.

15. Homeostatic regulation of the cardiovascular system ensures that each tissue receives adequate blood flow. This process is dependent most on:
    a. constant arterial blood pressure.
    b. constant venous blood pressure.
    c. ventricular systole.
    d. consistent heart rate.

16. Which of the following would interfere most with the ability of the veins to return blood to the heart?
    a. absence of valves in the veins
    b. low blood pressure in the veins
    c. high blood pressure in the arteries
    d. dilation of precapillary sphincters

17. The protective sac that encases the heart and anchors it to surrounding structures is the:
    a. pericardium.
    b. endocardium.
    c. endothelium.
    d. myocardium.

18. Atrioventricular valves prevent blood from flowing backward into the:
    a. atria.
    b. ventricles.
    c. pulmonary veins.
    d. aorta.

19. The movement of blood through the pulmonary circuit transports:
    a. oxygenated blood from the lungs directly to the body.
    b. oxygenated blood from the lungs to the heart.
    c. deoxygenated blood from the lungs to the heart.
    d. deoxygenated blood from the left side of the heart to the right side of the heart.

20. The ventricles fill with blood passively during:
    a. atrial systole.
    b. ventricular systole.
    c. atrial and ventricular diastole.
    d. semilunar diastole.

21. Which of the following represents the correct order in which electrical impulses will pass through the heart?
    a. AV node, AV bundle, SA node, Purkinje fibers
    b. SA node, Purkinje fibers, AV node, AV bundle
    c. AV bundle, AV node, SA node, Purkinje fibers
    d. SA node, AV node, AV bundle, Purkinje fibers

22. Which of the following would be occurring in a patient who lacked a P wave?
    a. no contraction of the atria
    b. no relaxation of the atria
    c. no contraction of the ventricles
    d. no relaxation of the ventricles

23. Shortness of breath, swollen ankles, and weight gain are symptoms of:
    a. hypertension.
    b. hypotension.
    c. cerebral embolism.
    d. congestive heart failure.

24. An increase in cardiac output would result from all of the following *except:*
    a. epinephrine.
    b. stimulation by sympathetic nerves.
    c. a decrease in stroke volume.
    d. an increase in heart rate.

25. If the cardiovascular center in the brain senses a drop in blood pressure, which of the following would occur?
    a. Heart rate would be lowered.
    b. Cardiac output would be reduced.
    c. Vasoconstriction of the arterioles would occur.
    d. Parasympathetic nerves would carry signals to the heart.

# Key Concept Review Questions

*Each of the Key Concepts listed at the beginning of this chapter has been rewritten as a question below. After successfully completing the study guide exercises and the Chapter Test, you should be able to answer each of these questions. Refer to the Key Concepts list to check your answers.*

1. What provides the power to transport blood?
2. Which vessels transport blood away from the heart? Which vessels transport blood toward the heart?
3. What is the function of the capillaries?
4. What is important about the structure of blood vessel walls?
5. What three mechanisms help veins return blood to the heart?
6. How does the lymphatic system help to maintain blood volume?
7. How would you describe the four chambers of the heart?
8. What is the function of the heart valves?
9. What type of blood enters and then leaves each side of the heart?
10. What is the function of the pulmonary circuit?
11. What is the function of the systemic circuit?
12. How would you describe the periods of the cardiac cycle?
13. What is the role of the cardiac conduction system?
14. What is blood pressure?
15. Why is blood pressure important?
16. What produces systolic pressure? What produces diastolic pressure?
17. What three variables must be regulated to maintain a constant arterial blood pressure?

18. What factors contribute to regulation of the cardiovascular system?

19. What are six disorders of the cardiovascular system discussed in Chapter 8?

20. How much effect can personal lifestyle choices have on cardiovascular health?

---

# Answer Key

## Sections 8.1, 8.2

**1.**p; **2.**v; **3.**r; **4.**a; **5.**t; **6.**j; **7.**b; **8.**g; **9.**e; **10.**o; **11.**x; **12.**s; **13.**c; **14.**w; **15.**d; **16.**n; **17.**u; **18.**m; **19.**h; **20.**f; **21.**y; **22.**l; **23.**k; **24.**i; **25.**q; **26.** arteries, capillaries, veins; **27.** pores, slits; **28.** aneurysm; **29.** vasoconstriction; **30.** Vasodilation; **31.** diffusion; **32.** Capillary beds; **33.** blood pressure; **34.** lymphatic capillaries; **35.** lymph; **36.** veins; **37.** varicose; **38.** pericardium; **39.** pericardial cavity; **40.** arrhythmia; **41.** endocarditis; **42.** T-wave; **43.** QRS complex; **44.** f,k; **45.**h; **46.**i,s; **47.**a,l,o,p; **48.**f,k; **49.**h; **50.**i,s; **51.**d,l,o,p; **52.**f,k; **53.**h,m,q,t; **54.**i,s; **55.**e; **56.**f,k; **57.**h,q; **58.**c,r; **59.**g; **60.**b,j,n; **61.** blood transport, blood storage, regulation of blood flow into the capillaries; **62.** capillary walls consist of a single layer of endothelium comprised of epithelial cells; **63.** low pressure in the veins; **64.** skeletal muscle squeezes the veins, valves prevent blood from flowing backward, pressure associated with breathing pushes blood toward the heart; **65.** sinoatrial node, atrioventricular node, atrioventricular bundle, Purkinje fibers; **66.** epicardium: outermost layer consisting of epithelial and connective tissue, myocardium: thick middle muscular layer responsible for contraction, endocardium: innermost layer of the heart continuous with the lining of the blood vessels; **67.** atria; **68.** ventricles; **69.** systole; **70.** diastole; **71.** diastole; **72.** atria; **73.** ventricles; **74.** atrial systole; **75.** ventricular systole; **76.** AV; **77.** pulmonary trunk; **78.** aorta; **79.** systole; **80.** diastole; **81.** pulse; **82.** heart sounds; **83.** conduction; **84.** SA node; **85.** atrium; **86.** AV node; **87.** AV bundle; **88.** brain; **89.** a. right atrioventricular valve, b. right ventricle, c. pulmonary semilunar valve, d. pulmonary trunk, e. lungs, f. pulmonary veins, g. left atrium; h. left atrioventricular valve, i. left ventricle, j. aortic semilunar valve, k. aorta, l. arteries, m. arterioles, n. capillaries, o. venules, p. veins; **90.** a. aorta, b. pulmonary semilunar valve, c. right atrium, d. right AV valve, e. right ventricle, f. pulmonary trunk, g. left atrium, h. aortic semilunar valve, i. left AV valve, j. left ventricle, k. septum

## Sections 8.3, 8.4, 8.5, 8.6

**1.**i; **2.**k; **3.**c; **4.**h; **5.**a; **6.**d; **7.**b; **8.**f; **9.**g; **10.**e; **11.** sphygmomanometer; **12.** systolic; **13.** isolated systolic hypertension; **14.** Hypotension; **15.** heart rate, stroke volume; **16.** Sympathetic, medulla oblongata; **17.** one; **18.** angiogram; **19.** CABG; **20.** greater distance from the heart causes blood pressure to dissipate; **21.** blood pressure is steady in the capillaries and pulsatile in the arteries; **22.** blood moves with the pressure gradient; **23.** arterial; **24.** aorta; **25.** carotid arteries; **26.** brain;

27. heart rate; 28. increase; 29. negative feedback; 30. medulla oblongata; 31. Sympathetic; 32. parasympathetic; 33. diameter; 34. epinephrine; 35. adrenal gland; 36. oxygen; 37. carbon dioxide; 38. dilate;
39. a. varied, b. often no symptoms; 40. a. decreased blood flow to the heart, b. pain and tightness in the chest; 41. a. death of heart tissue due to oxygen deprivation, b. intense chest pain, nausea, heaviness, difficulty breathing, radiating pain to left arm, back, or abdomen;
42. a. heart is less efficient at pumping blood due to age, prior heart attacks, deficient valves, hypertension, b. difficulty breathing, swollen legs and ankles, swollen neck veins, weight gain; 43. a. sudden blockage of a blood vessel by a clot or other obstruction, b. varied;
44. a. damage to brain cells due to decreased oxygen flow to the brain, b. sudden weakness or paralysis on one side of the body, difficulty speaking, impaired vision, nausea

## Chapter Test

1. increases, increasing; 2. thinner; 3. larger; 4. less than; 5. less than; 6. increases; 7. deoxygenated; 8.d; 9.c; 10.a; 11.b; 12.c; 13.c; 14.c; 15.a; 16.a; 17.a; 18.a; 19.b; 20.c; 21.d; 22.a; 23.d; 24.c; 25.c

# 9

# The Immune System and Mechanisms of Defense

## Chapter Summary and Key Concepts

*After reading and studying this chapter you should know the following:*

### Sections 9.1, 9.2, 9.3, 9.4, 9.5

1. The immune system is a group of cells, proteins, and structures of the lymphatic and circulatory systems that defend the body.

2. Pathogens are microorganisms that cause cell death and disease.

3. The danger of a particular pathogen is determined by its transmissibility, mode of transmission, and virulence.

4. The lymphatic system consists of a series of vessels, nodes, and organs that transport and cleanse the lymph.

5. The three lines of defense of the body are physical and chemical barriers, nonspecific defense mechanisms, and specific defense mechanisms.

6. The first line of defense, consisting of physical and chemical barriers, includes skin, tears, saliva, earwax, digestive acids, mucus, vomiting, urination, defecation, and resident bacteria.

7. The second line of defense consists of nonspecific mechanisms that attack foreign and damaged cells, including phagocytosis by neutrophils and macrophages, destruction of large invaders by eosinophils, the inflammatory response, natural killer cells, the complement system, interferons, and fever.

8. The third line of defense involves specific mechanisms that target specific pathogens for destruction, and depends on B cells and T cells.

### Sections 9.6, 9.7, 9.8, 9.9

9. The three important characteristics of the immune response are (1) it recognizes and targets specific pathogens, (2) it has memory, and (3) it protects the entire body.

10. An immune response is any response utilizing the components of the immune system. An antigen is any substance that can provoke an immune response.

11. Body cells are normally protected from the immune response by the presence of major histocompatibility complex (MHC) proteins.

12. Lymphocytes called B cells are responsible for antibody-mediated immunity. B cells produce antibodies capable of destroying a particular antigen.

13. Lymphocytes called T cells are responsible for cell-mediated immunity. T cells exist in several types and function to stimulate other cells and destroy infected body cells.

14. B cells recognize and bind to free antigens. T cells can only recognize and bind to an antigen that is presented by an antigen-presenting cell.

15. Both B cells and T cells produce memory cells that trigger the immune response faster when activated a second time. Memory cells are responsible for immunity.

16. Active immunization occurs when the body is exposed to a vaccine.

17. Passive immunization occurs when a person receives prepared antibodies for a specific pathogen.

18. Antibiotics are only effective against bacteria.

**Sections 9.10, 9.11**

19. Disorders of the immune system include allergies, lupus erythematosus, rheumatoid arthritis, and AIDS.

20. AIDS results from infection by the human immunodeficiency virus (HIV), and it causes a reduction in the number of helper T cells.

# Exercises

*Complete the exercises for each section after you have read and studied the section. If you cannot answer some questions, or answer them incorrectly, return to the chapter and review this information. You may find it helpful to work on only one section at a time. When you have completed all sections, take the Chapter Test as an indicator of your mastery of this topic.*

9.1 **Pathogens cause disease**

9.2 **The lymphatic system defends the body**

9.3 **The body has three lines of defense**

9.4 **Physical and chemical barriers: Our first line of defense**

9.5 **Nonspecific defenses: The second line of defense**

## Matching

____ 1. **immune system**
____ 2. **pathogens**
____ 3. **bacteria**
____ 4. **virus**
____ 5. **lymphatic system**
____ 6. **lymph nodes**
____ 7. **spleen**
____ 8. **thymus gland**
____ 9. **lysozyme**
____ 10. **phagocytosis**
____ 11. **neutrophils**
____ 12. **macrophages**
____ 13. **eosinophils**
____ 14. **inflammation**
____ 15. **mast cells**
____ 16. **histamine**
____ 17. **basophils**
____ 18. **natural killer cells**
____ 19. **complement system**
____ 20. **interferon**
____ 21. **fever**

a. the largest lymphatic organ, responsible for removing old red blood cells from circulation and fighting infections
b. a group of white blood cells that is nonspecific in destroying tumor cells and infected body cells
c. abnormally high body temperature
d. white blood cells that cluster around large invaders and destroy them by releasing digestive enzymes
e. a process in which white blood cells engulf and destroy foreign cells
f. connective tissue cells stimulated by the inflammatory response to release histamine
g. a group of cells, proteins, and structures of the lymphatic and circulatory systems that defend the body
h. prokaryotic organisms
i. the first white blood cells to respond to an infection
j. a group of 20 plasma proteins that assist defense mechanisms
k. monocytes that develop into large phagocytic cells
l. body system responsible for maintaining blood volume, transporting fats, and defending the body
m. site of maturation of T cells
n. white blood cells that secrete histamine
o. an enzyme found in tears and saliva that kills many bacteria
p. small organs of the lymphatic system that remove microorganisms, debris, and abnormal cells from the lymph
q. the smallest known infectious agent
r. microorganisms that cause cell death and disease
s. a protein secreted by virus-infected cells that diffuses to healthy neighboring cells and aids protection
t. a chemical that promotes vasodilation of small blood vessels
u. a response to tissue injury that involves redness, warmth, swelling, and pain

## Short Answer

22. What are the three functions of the lymphatic system?

23. Why are lymphatic capillaries able to absorb large substances that cannot enter a blood capillary?

Chapter 9  The Immune System and Mechanisms of Defense   125

24. What are the two functions of the spleen?

25. What are the three lines of defense that help to protect the body?

26. List the sequence of events involved in the inflammatory response.

## Labeling

*Label the indicated structures in Figure 9.1. Indicate both (a) the structure, and (b) the function.*

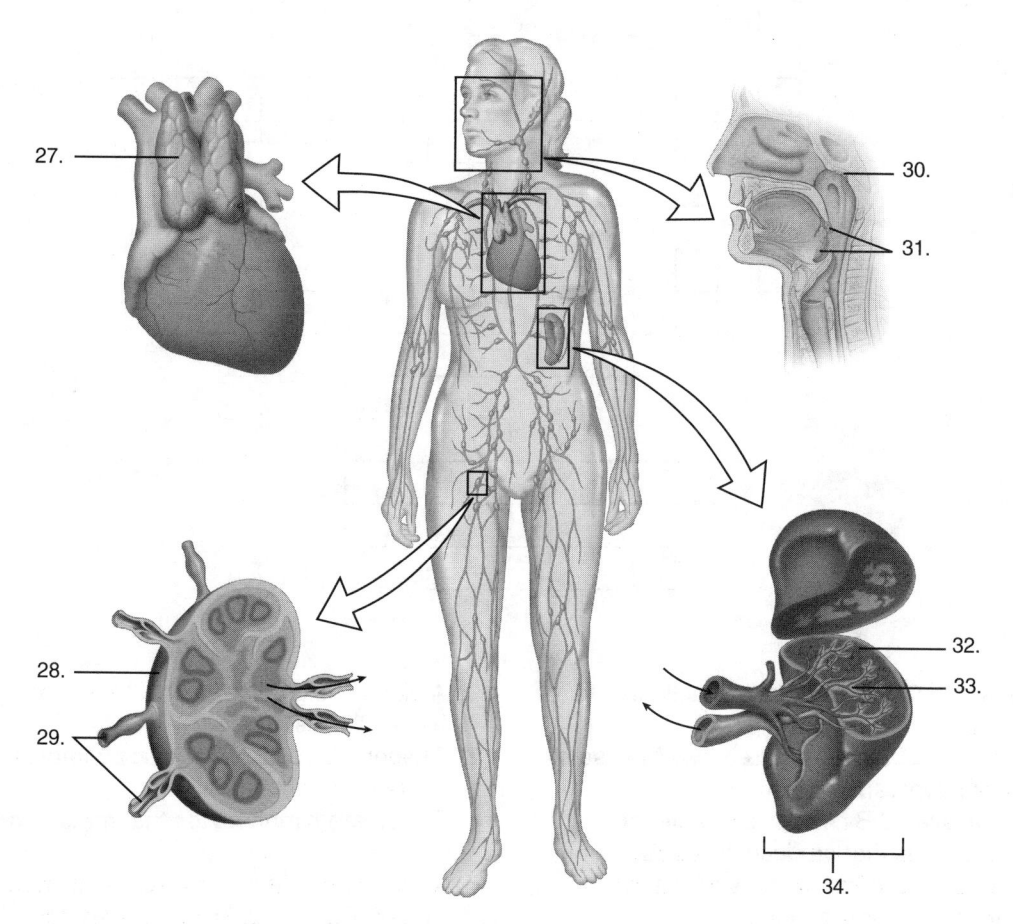

**Figure 9.1**

27. a. _____
    b. _____
28. a. _____
    b. _____
29. a. _____
    b. _____

30. a. _____
    b. _____
31. a. _____
    b. _____
32. a. _____
    b. _____

33. a. _____
    b. _____
34. a. _____
    b. _____

9.6 Specific defense mechanisms: The third line of defense
9.7 Immune memory creates immunity
9.8 Medical assistance in the war against pathogens
9.9 Tissue rejection: A medical challenge

## Crossword Puzzle

### Across

1. T cells with CD8 receptors are activated to produce _____ T cells
3. Tissue _____ is minimized with the use of immunosuppresive drugs
7. Some activated B cells differentiate into _____ cells that produce antibodies
9. _____ antibodies are pure preparations specific for a single antigen
11. Self-markers on body cells are known as _____ histocompatability complex proteins
12. The region of an antibody where the amino acid sequence varies
13. A protein released by cytotoxic T cells that makes holes in the target cell membrane
14. Both B cells and cytotoxic T cells require stimulation by _____ T cells

### Down

1. A class of signaling molecules secreted by helper T cells
2. Antigen-_____ cells display fragments of antigen on their surface
4. T cells are lymphocytes that mature in this gland
5. A first exposure to a particular antigen generates a _____ immune response
6. Antibiotics are effective against these microorganisms
8. B cells are responsible for _____-mediated immunity
10. Active immunization occurs when this is administered
11. The presence of these cells is the basis of immunity

## Ordering

*Place the following events pertaining to B-cell function in the proper sequence, beginning with "1" and ending with "5."*

_____ 15. activation of B-cell causes cloning of the B-cell

_____ 16. plasma cells produce antibodies

_____ 17. B-cell encounters a foreign antigen

_____ 18. B-cells differentiate into plasma cells and memory cells

_____ 19. antibodies bind foreign antigen

## Fill-In

*Fill-in the function of each of the following:*

20. APC:

21. CD4 receptor:

22. interleukin:

23. helper T-cell:

24. cytotoxic T-cell:

25. perforin:

## Completion

Write the name of each type of antibody described in the table below. Indicate what percentage of immunoglobulins each antibody type comprises.

| Type of Antibody | % of Immunoglobulins | Function of Antibody |
|---|---|---|
| 26. | 27. | enters areas of the body covered by mucous membranes |
| 28. | 29. | activates the inflammatory response, involved in allergies |
| 30. | 31. | long-lived antibodies, can cross the placenta |
| 32. | 33. | function unclear, may activate B cells |
| 34. | 35. | first antibodies released during an immune response |

**9.10 Inappropriate immune system activity causes problems**

**9.11 Immune deficiency: The special case of AIDS**

## Fill-in-the-Blank

*Referenced sections are in parentheses.*

1. _____ is an inappropriate response of the immune system to a nondangerous pathogen. (9.10)

2. Anaphylactic shock is a severe _____ allergic reaction. (9.10)

3. _____ disorders result when the immune system targets "self" cells. (9.10)

4. Phase _____ of AIDS is characterized by a gradual decline in the number of T cells and an increased vulnerability to opportunistic infections. (9.11)

## Labeling

*Use the steps of HIV infection listed below to label the steps in Figure 9.2.*

a. the host cell makes a second strand of DNA to complement the first single strand of DNA

b. HIV enters the host cell

c. the host cell expresses the HIV genes and makes new viral RNA and proteins

d. using HIV enzymes, the host cell makes a single-stranded DNA copy of the viral RNA strand

e. the double-stranded DNA fragment is inserted into the host DNA

f. new HIV particles leave the host cell

g. HIV binds to a host cell CD4 receptor

5. _____

6. _____

7. _____

8. _____

9. _____

10. _____

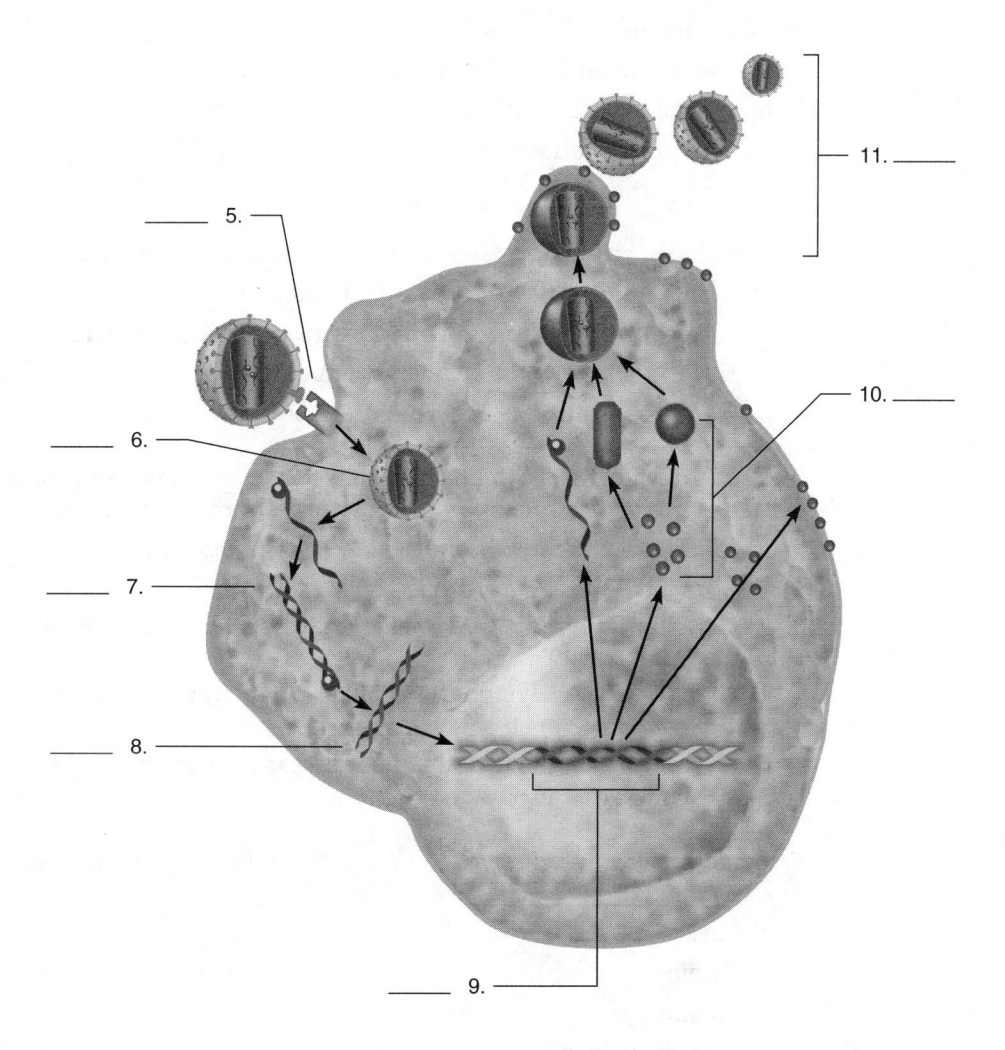

11. _____

**Figure 9.2**

# Chapter Test

## Multiple Choice

1. Microorganisms that cause cell death and disease are called:
   a. antigens.
   b. antibodies.
   c. macrophages.
   d. pathogens.

2. Characteristics of bacteria include all of the following *except:*
   a. presence of membrane-bound organelles.
   b. a single circular chromosome.
   c. absence of a nuclear membrane.
   d. presence of a cell wall.

3. Both viruses and prions:
   a. cause misfolding of proteins.
   b. contain DNA or RNA.
   c. are prokaryotic.
   d. are pathogenic.

4. Lymphatic vessels are similar to veins because they both:
   a. allow fluid to flow in two directions.
   b. use skeletal muscle contractions to keep fluids moving.
   c. have a single, unbranched structure.
   d. have the ability to absorb large molecules.

5. The correct path of lymph fluid collected from the tissues and transported to the heart is most accurately represented as:
   a. lymphatic duct → lymph node → lymphatic capillary → lymphatic vessel
   b. lymphatic vessel → lymphatic capillary → lymphatic duct → lymph node
   c. lymph node → lymphatic capillary → lymphatic vessel → lymphatic duct
   d. lymphatic capillary → lymphatic vessel → lymph node → lymphatic duct

6. Lymph nodes are clustered in all of the following areas of the body *except* the:
   a. neck.
   b. groin.
   c. ankles.
   d. digestive tract.

7. The spleen:
    a. is the largest lymphatic organ.
    b. includes white pulp that contains macrophages.
    c. cleanses the lymph.
    d. is required to maintain life and health.

8. Specific defense mechanisms differ from nonspecific defense mechanisms because only specific mechanisms:
    a. utilize macrophages.
    b. target particular antigens.
    c. protect a localized region.
    d. involve complement proteins.

9. What do lysozyme, dermicidin, and mucus have in common?
    a. All are produced by lymphocytes.
    b. All circulate in lymphatic fluid.
    c. They are all part of the first line of defense.
    d. They are all associated with the action of macrophages.

10. The first white blood cells to respond to an infection are:
    a. eosinophils.
    b. macrophages.
    c. neutrophils.
    d. helper B cells.

11. A group of at least 20 proteins that circulate in the blood and assist defense mechanisms is called:
    a. natural killer cells.
    b. interferons.
    c. complement.
    d. cytokines.

12. Fever is caused when _____ release _____.
    a. macrophages, pyrogens
    b. infected cells, interferon
    c. B cells, cytokines
    d. eosinophils, histamines

13. A substance capable of provoking an immune response is a(n):
    a. antibody.
    b. antigen.
    c. lymphokine.
    d. complement.

14. Major histocompatibility complex proteins:
    a. are found in the cytoplasm.
    b. attract pathogens.
    c. identify a cell as "self."
    d. are produced by B cells.

15. B cells mature in the _____, and T cells mature in the _____.
    a. bone marrow, bone marrow
    b. bone marrow, thymus gland
    c. thymus gland, bone marrow
    d. thymus gland, thymus gland

16. Specific immunity is dependent on the body's ability to:
    a. distinguish its own cells from those of foreign invaders.
    b. manufacture adequate numbers of red blood cells.
    c. maintain a first and second line of defense.
    d. circulate and replenish lymphatic fluid.

17. B-cells are responsible for _____ immunity, and T-cells are responsible for _____ immunity.
    a. cell-mediated, antibody-mediated
    b. antibody-mediated, cell-mediated
    c. specific, nonspecific
    d. active, passive

18. When a T cell with CD8 receptors contacts an APC displaying an antigen-MHC complex:
    a. the APC divides to form helper T cells.
    b. the T cell produces a clone of cytotoxic T cells.
    c. the antigen-MHC complex is destroyed.
    d. the CD8 receptors are converted to CD4 receptors.

19. A specific antibody will bind to:
    a. a general class of antigens.
    b. a specific antigen.
    c. any antigen because the heavy chains can accommodate any shape.
    d. any antigen because the binding sites are variable.

20. Which of the following can function as antigen-presenting cells?
    a. natural killer cells
    b. macrophages
    c. mast cells
    d. antibodies

21. A secondary immune response could not occur without:
    a. memory cells.
    b. complement.
    c. neutrophils.
    d. natural killer cells.

22. Administration of prepared antibodies produces:
    a. active immunity.
    b. passive immunity.
    c. delayed immunity.
    d. permanent immunity.

23. The actions and effectiveness of antibiotics depends on differences between bacterial and human cells, including:
    a. bacteria have a thick cell wall, human cells do not.
    b. the membrane-bound organelles of bacteria contain different proteins than those of human cells.
    c. bacteria do not have ribosomes.
    d. human protein synthesis occurs faster than bacterial protein synthesis.

24. HIV interferes with specific immunity because it infects and eventually destroys:
    a. B cells.
    b. cytotoxic T cells.
    c. helper T cells.
    d. MHC cells.

25. Autoimmune disorders occur when:
    a. the immune system loses a large number of T cells.
    b. the ability of B cells to produce antibodies is reduced.
    c. the immune system fails to recognize "self" cells.
    d. macrophages function only in nonspecific immunity.

# Key Concept Review Questions

*Each of the Key Concepts listed at the beginning of this chapter has been rewritten as a question below. After successfully completing the study guide exercises and the Chapter Test, you should be able to answer each of these questions. Refer to the Key Concepts list at the beginning of this chapter to check your answers.*

1. The immune system comprises which molecules, structures, and systems?

2. What are pathogens?

3. What determines the danger of a particular pathogen?

4. The lymphatic system comprises which structures and organs?

5. What are the three lines of defense of the body?

6. Which structures and processes make up the first line of defense?

7. How would you describe the structures and processes that make up the second line of defense?

8. Is the third line of defense specific or nonspecific? What cells are responsible for the third line of defense?

9. What are the three important characteristics of the immune response?

10. What is an immune response? What is an antigen?

11. How are body cells normally protected from the immune system?

12. What type of immunity involves B cells? What molecules do B cells produce?

13. What type of immunity involves T cells? What do T cells do?

14. What type of antigen is recognized by B cells? What type of antigen is recognized by T cells?

15. What process are memory cells responsible for? What cells produce memory cells?

16. What results in active immunity?

17. What results in passive immunity?

18. What pathogens are antibiotics effective against?

19. What are four disorders of the immune system discussed in the chapter?

20. What causes AIDS? What effect does AIDS have on T cells?

## Answer Key

### Sections 9.1, 9.2, 9.3, 9.4, 9.5

**1.**g; **2.**r; **3.**h; **4.**q; **5.**l; **6.**p; **7.**a; **8.**m; **9.**o; **10.**e; **11.**i; **12.**k; **13.**d; **14.**u; **15.**f; **16.**t; **17.**n; **18.**b; **19.**j; **20.**s; **21.**c; **22.** transportation, maintenance of blood volume, defense of the body; **23.** the spaces between the overlapping cells of lymphatic capillaries are large enough for passage of substances such as bacteria; **24.** removing old and damaged red blood cells, and participation in defense of the body; **25.** physical and chemical barriers, nonspecific defenses, specific defenses; **26.** damaged cells release chemicals that stimulate mast cells, basophils and mast cells release histamine which causes dilation of blood vessels, dilated blood vessels allow phagocytes to move into the interstitial fluid, and phagocytes attack pathogens and damaged cells; **27.** a. thymus gland, b. aids maturation of T cells; **28.** a. lymph nodes, b. filters and cleanses lymph; **29.** a. lymph vessels, b. transport lymph; **30.** a. adenoids, b. protect the throat; **31.** a. tonsils, b. protect the throat; **32.** a. red pulp, b. contains macrophages that break down microorganisms; **33.** a. white pulp, b. contains lymphocytes that fight infections; **34.** a. spleen, b. removes damaged red blood cells and helps to fight infections

## Sections 9.6, 9.7, 9.8, 9.9

**Crossword Puzzle: 1.** across: cytotoxic; down: cytokines; **2.** presenting; **3.** rejection; **4.** thymus; **5.** primary; **6.** bacteria; **7.** plasma; **8.** antibody; **9.** Monoclonal; **10.** vaccine; **11.** across: major; down: memory; **12.** variable; **13.** perforin; **14.** helper; **15.** 2; **16.** 4; **17.** 1; **18.** 3; **19.** 5; **20.** to engulf a pathogen and display antigen-MHC complex on the cell surface; **21.** is found on T-cells that will develop into helper T-cells; **22.** a cytokine molecule that helps to promote the development of other immune cells; **23.** stimulates the development of other immune cells; **24.** destroys infected body cells; **25.** protein molecule that is released by cytotoxic T-cells, attaches to the cell membrane of target infected body cells, and causes lysis of the target cell; **26.** IgA; **27.** 15%; **28.** IgE; **29.** 0.1%; **30.** IgG; **31.** 75%; **32.** IgD; **33.** less than 1%; **34.** IgM; **35.** 5–10%

## Sections 9.10, 9.11

**1.** Allergy; **2.** systemic; **3.** Autoimmune; **4.** II; **5.** g; **6.** b; **7.** d; **8.** a; **9.** e; **10.** c; **11.** f

## Chapter Test

**1.** d; **2.** a; **3.** d; **4.** b; **5.** d; **6.** c; **7.** a; **8.** a; **9.** c; **10.** a; **11.** c; **12.** a; **13.** b; **14.** c; **15.** b; **16.** a; **17.** b; **18.** b; **19.** b; **20.** b; **21.** a; **22.** b; **23.** a; **24.** c; **25.** c

# 10

# The Respiratory System: Exchange of Gases

## Chapter Summary and Key Concepts

*After reading and studying this chapter you should know the following:*

**Sections 10.1, 10.2**

1. The term respiration includes four processes: ventilation, external respiration, internal respiration, and cellular respiration.

2. Ventilation is the process of moving air into and out of the lungs, external respiration is the exchange of gases between inhaled air and the blood, internal respiration is the exchange of gases between the blood and tissue fluids, and cellular respiration is the process of using oxygen to produce ATP within cells.

3. The upper respiratory tract consists of the nose and pharynx, and it functions to filter, warm, and humidify air.

4. The lower respiratory tract consists of the larynx, trachea, bronchi, and lungs, and it functions to move air and exchange gases.

5. The pharynx is a common passageway for food, liquid, and air. Food and liquid are prevented from moving into the air passageways by the epiglottis.

6. The larynx also functions as the voice box. Vibration of the vocal cords produces sound.

7. The trachea is reinforced with incomplete cartilage rings; it branches into the bronchi, which are reinforced by cartilage tissue; the bronchi further branch into the bronchioles, which have no cartilage.

8. Bronchioles terminate in tiny air-filled sacs called alveoli that function as the site of gas exchange.

9. The lungs are in the thoracic cavity and are enclosed by the pleural membranes.

**Sections 10.3, 10.4, 10.5, 10.6**

10. Contraction of the intercostal muscles and the diaphragm enlarge the thoracic cavity, resulting in enlargement of the lungs.

11. Air pressure within the lungs decreases when lung volume increases.

12. Air moves into the lungs down a pressure gradient.

13. Inspiration is the movement of air into the lungs. It is an active process because it requires muscular effort.

14. Passive relaxation of the intercostals and the diaphragm reduces lung volume, increasing air pressure within the lungs. As the air pressure within the lungs becomes greater than the atmospheric pressure, air rushes out of the lungs.

15. Expiration is the movement of air out of the lungs. It is usually a passive process.

16. Tidal volume, vital capacity, inspiratory reserve volume, expiratory reserve volume, and residual volume are measurements of lung capacity and can reflect lung function.

17. External respiration and internal respiration occur when gases move passively along a pressure gradient.

18. Most oxygen is transported through the blood bound to hemoglobin. Carbon dioxide is transported through the blood bound to hemoglobin or in the form of a bicarbonate ion.

19. Breathing is regulated by the nervous system.

20. Disorders of the respiratory system include asthma, emphysema, bronchitis, pneumonia, tuberculosis, botulism, cancer, and congestive heart failure.

# Exercises

*Complete the exercises for each section after you have read and studied the section. If you cannot answer some questions, or answer them incorrectly, return to the chapter and review this information. You may find it helpful to work on only one section at a time. When you have completed all sections, take the Chapter Test as an indicator of your mastery of this topic.*

**10.1 Respiration takes place throughout the body**

**10.2 The respiratory system consists of upper and lower respiratory tracts**

## Matching

_____ 1. **ventilation**       a. the smallest airways

_____ 2. **nasal cavity**      b. two folds of connective tissue that extend across the larynx

_____ 3. **pharynx**           c. tiny air-filled sacs in the lungs that carry out gas exchange

_____ 4. **larynx**            d. organs of gas exchange

_____ 5. **epiglottis**        e. two layers of epithelial membrane enclosing each lung

_____ 6. **vocal cords**       f. the movement of air into and out of the lungs

_____ 7. **glottis**           g. airways that branch from the trachea and enter the lungs

_____ 8. **trachea**           h. the opening into the larynx

_____ 9. **bronchi**           i. the internal portion of the nose

_____ 10. **bronchioles**      j. the voice box

_____ 11. **lungs**            k. the "windpipe" extending from the larynx to the bronchi

_____ 12. **pleural membranes**  l. the throat

_____ 13. **alveoli**          m. a flexible cartilage flap located at the opening to the larynx

## Fill-in-the-Blank

*Referenced sections are in parentheses.*

14. The four processes included in respiration are _____, _____, _____, and _____. (10.1)

15. The exchange of gases between the blood and tissue fluids is _____ respiration. (10.1)

16. The process of using oxygen to produce ATP within cells is _____ respiration. (10.1)

17. In addition to gas exchange, the respiratory system is also responsible for the production of _____. (10.1)

18. The upper respiratory tract consists of the _____ and the _____. (10.2)

19. The lower respiratory tract consists of the _____, _____, _____, and _____. (10.2)

20. The two chambers of the nose are separated by the _____ _____. (10.2)

21. Gas exchange between the air in the alveoli and the blood in the pulmonary capillaries occurs by _____. (10.2)

22. Both food and air pass through the _____. (10.2)

23. The pharynx connects the mouth and nasal cavity to the _____. (10.2)

24. _____ are air spaces in the skull that secrete mucus and drain into the nasal cavity. (10.2)

25. The larynx contains the _____ and the _____. (10.2)

26. The _____ are marked by the absence of cartilage. (10.2)

27. Friction in the pleural membrane during lung movement is reduced by the presence of _____ in the _____ _____. (10.2)

28. The right lung has _____ lobes and the left lung has _____ lobes. (10.3)

29. The wall of a single alveolus is _____ layer(s) of _____ _____ cells. (10.2)

30. Surface tension within the alveoli is reduced by _____. (10.2)

31. Gas exchange in the lungs occurs between the _____ and the _____ _____. (10.2)

140 Study Guide for *Human Biology*

## Labeling

*Label each of the structures indicated in Figure 10.1.*

32. _____

33. _____

34. _____

35. _____

36. _____

37. _____

38. _____

39. _____

40. _____

41. _____

42. _____

43. _____

44. _____

45. _____

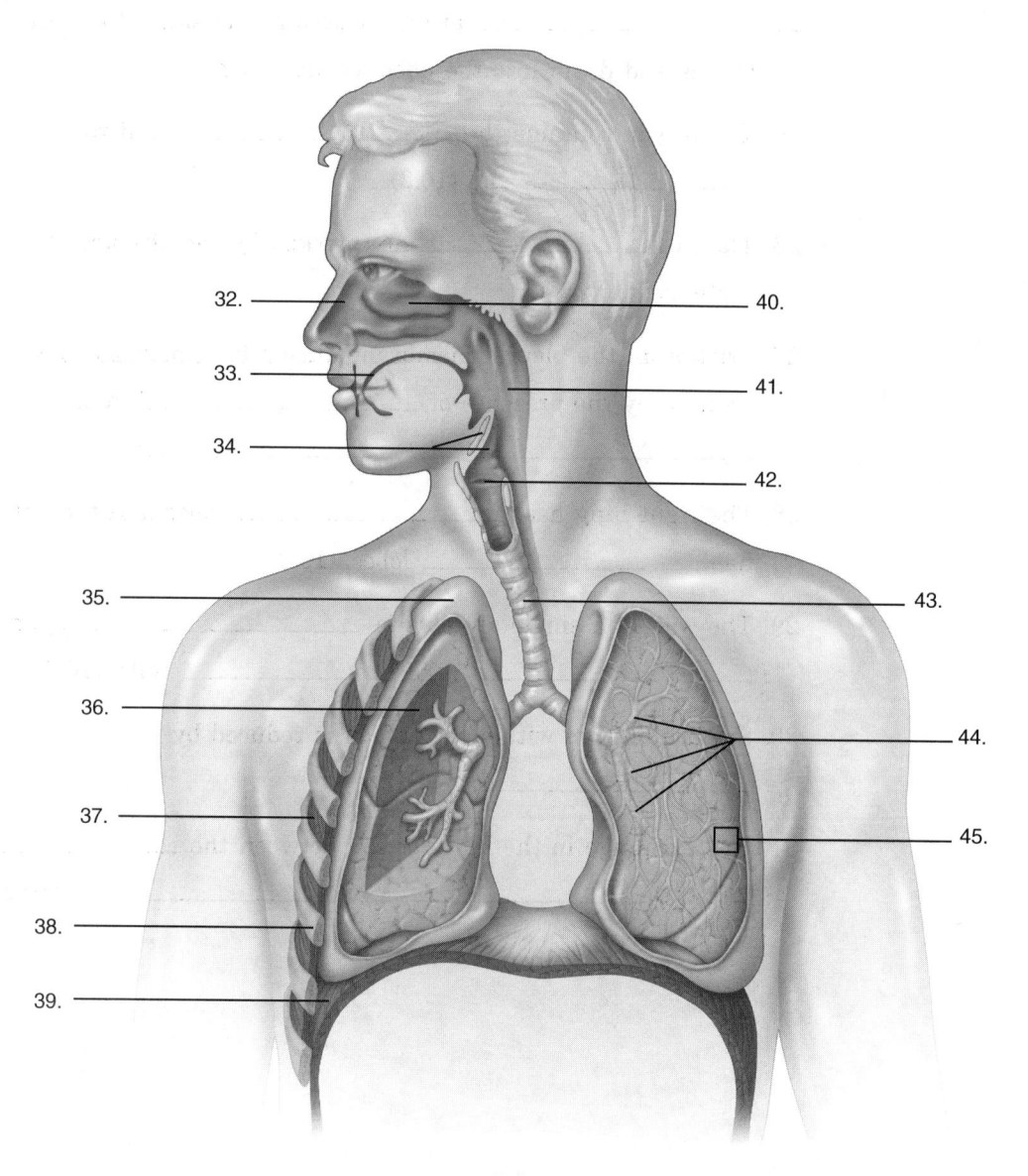

**Figure 10.1**

## 10.3 The process of breathing involves a pressure gradient

## Fill-in-the-Blank

1. Gas pressure is caused when gas molecules _____.

2. When the volume of a closed space increases, the gas pressure _____.

3. Gases flow from an area of _____ pressure to an area of _____ pressure.

4. Inspiration and expiration occur when air moves down its _____ _____.

## Short Answer

*Refer to Figure 10.2 to answer the questions below.*

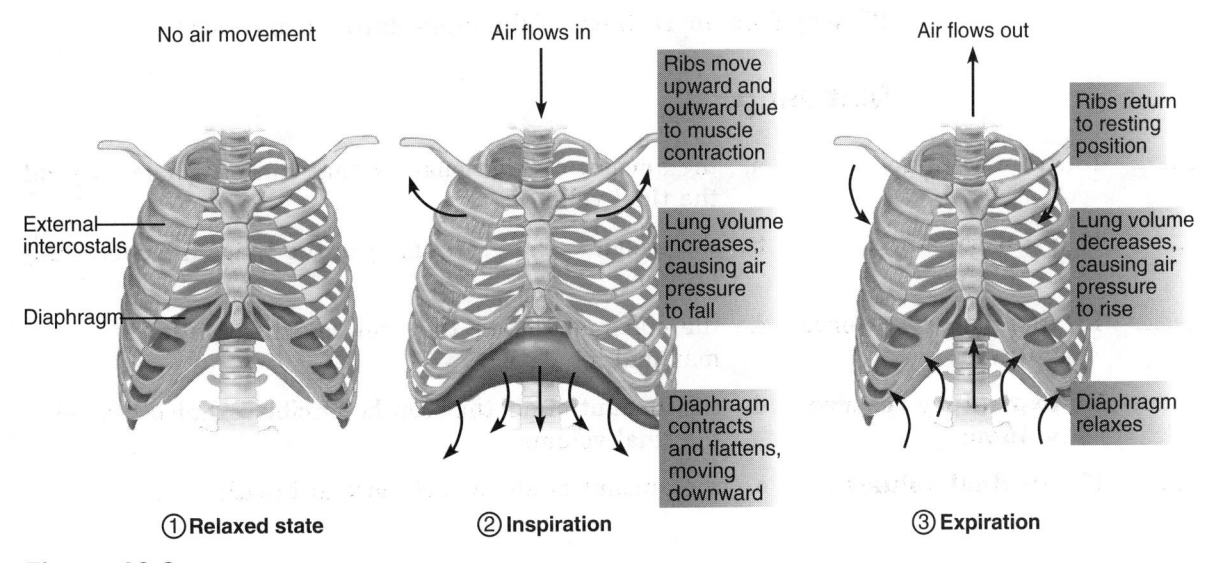

**Figure 10.2**

5. What muscles are involved in moving air into and out of the lungs?

6. What muscle activity is responsible for inspiration?

7. During inspiration, what change occurs in the volume of the lungs?

8. During inspiration, what change occurs in the air pressure in the lungs?

9. Why does air rush into the lungs during inspiration?

10. As the muscles of respiration relax, what change occurs in lung volume?

11. As the muscles of respiration relax, what change occurs in air pressure in the lungs?

12. Why does air rush out of the lungs during expiration?

## Matching

_____ 13. **tidal volume**   a. the amount of additional air that can be inhaled beyond the tidal volume

_____ 14. **vital capacity**   b. the amount of air remaining in the lungs after forcefully exhaling

_____ 15. **inspiratory reserve volume**   c. the maximal volume that can be exhaled after a maximal inhalation

_____ 16. **expiratory reserve volume**   d. the amount of air that can be forcibly exhaled beyond the tidal volume

_____ 17. **residual volume**   e. the amount of air in each normal breath

## 10.4 Gas exchange and transport occurs passively

## Short Answer

*Figure 10.3 depicts the partial pressure of oxygen and carbon dioxide in various areas of the body. Differences in partial pressure of a gas create a pressure gradient that drives the movement of the gas. Refer to Figure 10.3 to answer the following questions.*

**Breathing**

Dry inhaled air
$O_2$ 160
$CO_2$ 0.3

Moist exhaled air
$O_2$ 120
$CO_2$ 27

**Pulmonary circulation**

Lung capillaries

Alveolar air
$O_2$ 104
$CO_2$ 40

Alveolus 40 | 104
$CO_2$ 46 | $O_2$ 100
Capillary

**External respiration**

**Transport**

Pulmonary vein and aorta
$O_2$ 100
$CO_2$ 40

Systemic veins and pulmonary artery
$O_2$ 40
$CO_2$ 46

**Internal respiration**

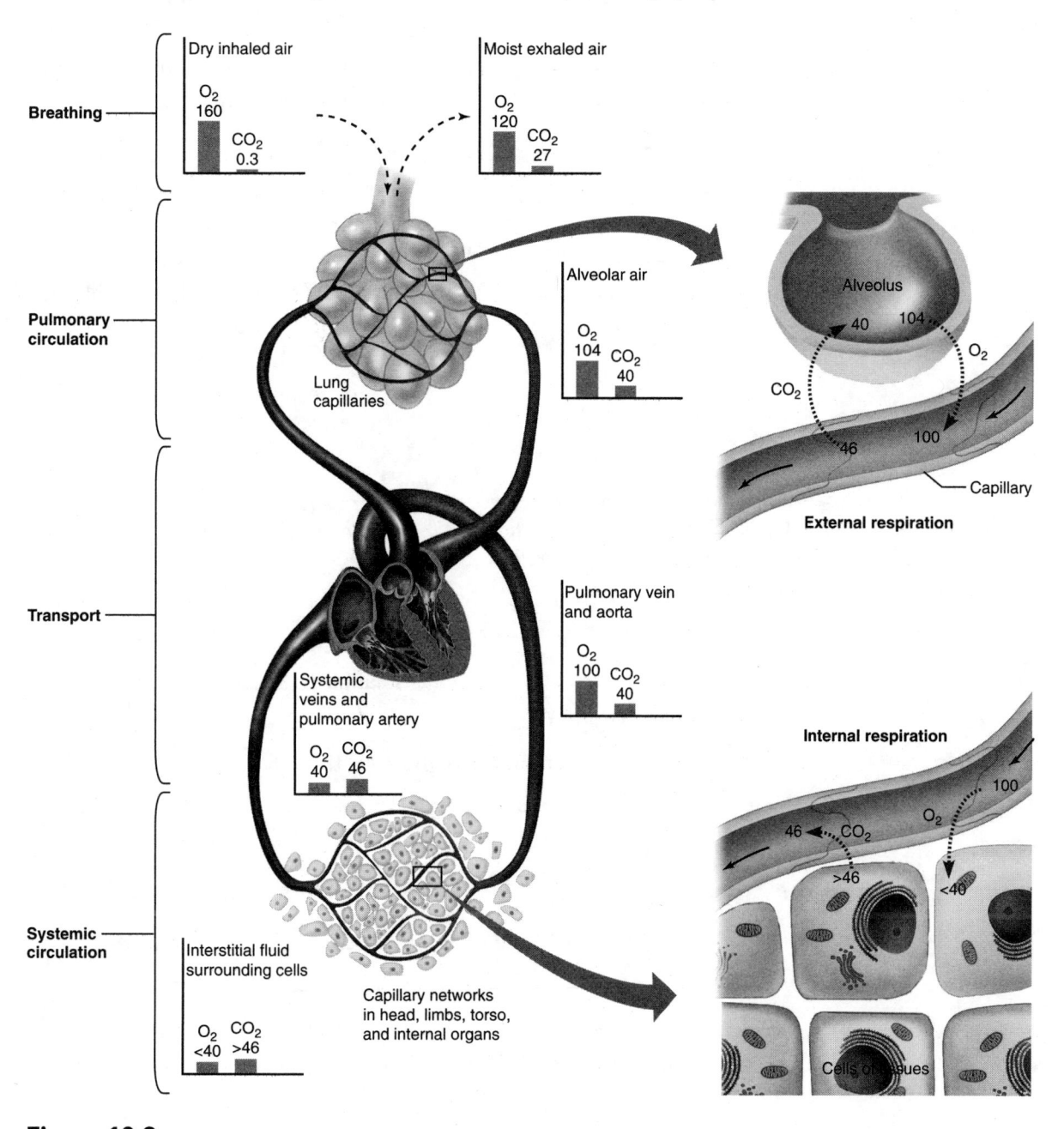

$CO_2$ 46 | $O_2$ 100
>46 | <40

**Systemic circulation**

Interstitial fluid surrounding cells
$O_2$ <40
$CO_2$ >46

Capillary networks in head, limbs, torso, and internal organs

Cells of tissues

**Figure 10.3**

1. Gas exchange between the alveoli and the pulmonary capillaries:
   _____ a. What is the $PO_2$ in the alveoli?
   _____ b. What is the $PO_2$ in the pulmonary capillaries?
   _____ c. What direction will $O_2$ move?
   _____ d. What is the $PCO_2$ in the alveoli?
   _____ e. What is the $PCO_2$ in the pulmonary capillaries?
   _____ f. What direction will $CO_2$ move?
   _____ g. What type of respiration is this?

2. Gas exchange between the capillary blood and the interstitial fluid:
   _____ a. What is the $PO_2$ in the capillary blood?
   _____ b. What is the $PO_2$ in a cell?
   _____ c. What direction will $O_2$ move?
   _____ d. What is the $PCO_2$ in the capillary blood?
   _____ e. What is the $PCO_2$ in a cell?
   _____ f. What direction will $CO_2$ move?
   _____ g. What type of respiration is this?

## Fill-in-the-Blank

*Each formula below represents a process involved in the transport of $O_2$ or $CO_2$ in the blood. Answer the questions about each process below the formula for that process.*

3. **Hb + $O_2$ → Hb$O_2$**
   a. Hb$O_2$ is a molecule of _____ bound to _____ and is called _____.
   b. This reaction is _____.
   c. List three conditions that promote the binding of $O_2$ to hemoglobin: _____, _____, and _____.

4. **Hb + $CO_2$ → HbC$O_2$**
   a. HbC$O_2$ is a molecule of _____ bound to _____ _____ and is called _____.
   b. What percentage of carbon dioxide is transported in the blood this way? _____

5. $CO_2 + H_2O \rightarrow H_2CO_3 \rightarrow HCO_3 + H^+$

   a. In this reaction, $CO_2$ combines with water in the blood to produce a _____ ion.

   b. What percentage of carbon dioxide is transported in the blood this way? _____

### 10.5 The nervous system regulates breathing

### 10.6 Disorders of the respiratory system

## Matching

_____ 1. **respiratory center**  a. receptors in the arteries that respond to the $PO_2$ of the blood

_____ 2. **stretch receptors**  b. receptors in the lungs that monitor the degree of lung inflation

_____ 3. **cerebrospinal fluid**  c. an area of the brain that regulates the rate of breathing

_____ 4. **carotid and aortic bodies**  d. changes in the $H^+$ concentration here are detected by cells in the brain

## Short Answer

*For each respiratory system disorder listed below, indicate the cause and the symptoms.*

5. Asthma
   a. cause:

   b. symptoms:

6. Emphysema
   a. cause:

   b. symptoms:

7. Bronchitis
   a. cause:

   b. symptoms

8. Pneumonia
   a. cause:

   b. symptoms:

9. Tuberculosis
   a. cause:

   b. symptoms:

10. Botulism
    a. cause:

    b. symptoms:

11. Lung cancer
    a. cause:

    b. symptoms:

12. Cystic fibrosis
    a. cause:

    b. symptoms:

# Chapter Test

## Multiple Choice

1. The exchange of gases between inhaled air and the blood is called:
   a. ventilation.
   b. internal respiration.
   c. external respiration.
   d. cellular respiration.

2. The larynx, trachea, bronchi, and lungs make up the:
   a. respiratory system.
   b. upper respiratory tract.
   c. lower respiratory tract.
   d. unrelated respiratory organs.

3. Mucus in the nasal cavity functions to:
   a. maintain the required pH of nasal secretions.
   b. trap dust and pathogens in incoming air.
   c. guide airflow in the proper direction.
   d. enhance the production of cilia.

4. The functions of the nose include all of the following *except*:
   a. receptors provide for the sense of smell.
   b. gas exchange.
   c. filtering inhaled air.
   d. moistening air.

5. When food is traveling from the pharynx to the esophagus, the:
   a. glottis is open.
   b. epiglottis is open.
   c. epiglottis blocks the glottis.
   d. glottis blocks the epiglottis.

6. Which of the following structures that influence vocalization are incorrectly matched with their function?
   a. tongue and soft palate—create recognizable sounds
   b. pharynx and nose—act as resonating chambers to amplify and enhance vocal tone
   c. vocal cords—vibrate to produce sound
   d. nasal sinuses—push air through the vocal cords for sound production

7. The incomplete cartilage rings of the trachea allow the trachea to:
   a. maintain a constant diameter at all times.
   b. push air downward with muscle movements.
   c. close when needed to prevent choking.
   d. change diameter to accommodate coughing or heavy breathing.

8. The alveoli:
   a. consist of a single layer of squamous epithelial cells.
   b. are reinforced by cartilage tissue.
   c. do not participate in gas exchange.
   d. fuse to form the bronchioles.

9. The lungs are enclosed by a double pleural membrane. These membranes are attached to the _____ and the _____.
   a. thoracic cavity, pleural cavity
   b. pleural cavity, lung surface
   c. thoracic cavity, lung surface
   d. thoracic cavity, abdominal cavity

10. Gas exchange between inhaled air and the blood is accomplished by the _____ and the _____.
    a. alveoli, pulmonary veins
    b. bronchioles, pulmonary capillaries
    c. alveoli, pulmonary capillaries
    d. alveoli, bronchioles

11. Gases flow passively from an area of _____ pressure to an area of _____ pressure.
    a. high, low
    b. low, high
    c. high, high
    d. low, low

12. Air moves into the lungs when the air pressure in the lungs is _____ the air pressure in the atmosphere.
    a. lower than
    b. higher than
    c. the same as
    d. differences in air pressure are not responsible for airflow

13. Expiration is:
    a. always a passive process.
    b. always an active process.
    c. usually passive, but it can be active.
    d. usually active, but it can be passive.

14. The maximal volume of air that can be exhaled after a maximal inhalation is the:
    a. tidal volume.
    b. vital capacity.
    c. inspiratory reserve volume.
    d. expiratory reserve volume.

15. Measurements of lung capacity are performed with a device called a(n):
    a. volumetric analyzer.
    b. spirometer.
    c. inhalizer.
    d. airflow compression chamber.

16. Under which of the following conditions would oxygen flow from the alveoli into the blood?
    a. $PO_2$ in the alveoli is 40, $PO_2$ in the blood is 100
    b. $PO_2$ in the alveoli is 104, $PO_2$ in the blood is 104
    c. $PCO_2$ in the alveoli is 40, $PO_2$ in the alveoli is 46
    d. $PO_2$ in the alveoli is 104, $PO_2$ in the blood is 40

17. The earth's atmosphere consists primarily of three gases, which are:
    a. hydrogen, oxygen, and carbon.
    b. nitrogen, oxygen, and carbon dioxide.
    c. nitrogen, oxygen, and water.
    d. oxygen, hydrogen, and carbon dioxide.

18. Venous blood is defined as blood that is:
    a. found in the veins.
    b. low in carbon dioxide.
    c. traveling toward the heart.
    d. deoxygenated.

19. Hemoglobin in the blood transports:
    a. oxygen only.
    b. carbon dioxide only.
    c. oxygen and carbon dioxide in equal amounts.
    d. more oxygen than carbon dioxide.
    e. more carbon dioxide than oxygen.

20. Most of the carbon dioxide transported in the blood to the lungs will be transported:
    a. by hemoglobin.
    b. by red blood cells.
    c. as bicarbonate ions in the red blood cells.
    d. as bicarbonate ions in the plasma.

21. A low pH reduces the binding affinity of hemoglobin for oxygen, increasing the release of oxygen and its delivery to the tissues. What produces a condition of low pH in the red blood cells?
    a. conversion of $CO_2$ to bicarbonate ions
    b. conversion of $O_2$ to oxyhemoglobin
    c. conversion of $CO_2$ to carbaminohemoglobin
    d. conversion of bicarbonate ions to $CO_2$

22. The respiratory center:
    a. monitors the level of oxygen in the blood.
    b. contains stretch receptors that limit lung expansion through negative feedback.
    c. controls the rate of breathing.
    d. is the site of gas exchange between the air and the blood.

23. The concentration of $H^+$ in the cerebrospinal fluid rises when:
    a. the $PO_2$ of the arterial blood rises.
    b. the $PO_2$ of the arterial blood falls.
    c. the $PCO_2$ of the arterial blood rises.
    d. the $PCO_2$ of the arterial blood falls.

24. The carotid and aortic bodies are arterial receptors that respond to:
    a. a decrease in the $PO_2$ of arterial blood.
    b. an increase in the $PO_2$ of arterial blood.
    c. a decrease in the $PO_2$ of venous blood.
    d. an increase in the $PO_2$ of venous blood.

25. A disorder of the respiratory system involving permanent damage to the alveoli is:
    a. asthma.
    b. bronchitis.
    c. emphysema.
    d. pneumonia.

## Key Concept Review Questions

*Each of the Key Concepts listed at the beginning of this chapter has been rewritten as a question below. After successfully completing the study guide exercises and the Chapter Test, you should be able to answer each of these questions. Refer to the Key Concepts list at the beginning of this chapter to check your answers.*

1. What are the four processes encompassed by the term respiration?
2. How would you describe what occurs in each of the four processes listed above?
3. What are the components and function of the upper respiratory tract?
4. What are the components and function of the lower respiratory tract?
5. What acts as a common passageway for food, liquid, and air? What prevents food and liquid from moving into the airways?
6. Where and how is sound produced?
7. How would you compare the amount of cartilage present in the trachea, bronchi, and bronchioles?
8. What structures are located at the end of the bronchioles? What process occurs here?
9. What body cavity contains the lungs? What membrane encloses the lungs?
10. What are two skeletal muscles that contribute to lung function?
11. What happens to the air pressure in the lungs as the lung volume increases?
12. How would you compare the air pressure in the atmosphere with air pressure in the lungs when air flows into the lungs?
13. What is inspiration? Why is it considered an active process?
14. What muscle activity reduces lung volume? Compare the air pressure in the atmosphere with the air pressure in the lungs when air flows out of the lungs.
15. What is expiration? Is it active or passive?
16. What are five measurements of lung capacity?

Chapter 10 *The Respiratory System: Exchange of Gases* 151

17. Do gases move with or against a pressure gradient during external and internal respiration?

18. How is most oxygen transported in the blood? How is carbon dioxide transported in the blood?

19. What body system regulates breathing?

20. What are eight disorders of the respiratory system discussed in the chapter?

## Answer Key

### Sections 10.1, 10.2

**1.**f; **2.**i; **3.**l; **4.**j; **5.**m; **6.**b; **7.**h; **8.**k; **9.**g; **10.**a; **11.**d; **12.**e; **13.**c; **14.** breathing, external respiration, internal respiration, cellular respiration; **15.** internal; **16.** cellular; **17.** sound; **18.** nose, pharynx **19.** larynx, trachea, bronchi, lungs; **20.** nasal septum; **21.** diffusion; **22.** pharynx; **23.** larynx; **24.** Sinuses; **25.** epiglottis, vocal cords; **26.** bronchioles; **27.** fluid, pleural cavity; **28.** three, two; **29.** one, squamous epithelium; **30.** surfactant; **31.** alveoli, pulmonary capillaries; **32.** nose; **33.** mouth; **34.** epiglottis; **35.** pleural membranes; **36.** lung; **37.** intercostals muscle; **38.** rib; **39.** diaphragm; **40.** nasal cavity; **41.** pharynx; **42.** larynx; **43.** trachea; **44.** bronchi; **45.** alveoli

### Section 10.3

**1.** collide; **2.** decreases; **3.** high, low; **4.** pressure gradient; **5.** intercostals, diaphragm; **6.** intercostals lift the ribs upward, diaphragm flattens; **7.** increases; **8.** decreases; **9.** air pressure outside the body is greater than air pressure in the lungs; **10.** decreases; **11.** increases; **12.** air pressure in the lungs is greater than air pressure outside the body; **13.**e; **14.**c; **15.**a; **16.**d; **17.**b

### Section 10.4

**1.** a. 104, b. 100, c. into the pulmonary capillaries, d. 40, e. 46, f. into the alveoli, g. external; **2.** a. 100, b. <40, c. into the cell, d. 46, e. >46, f. into the capillary, g. internal; **3.** a. hemoglobin, oxygen, oxyhemoglobin, b. reversible, c. neutral pH, cool temperature, high $PO_2$ in the lungs; **4.** a. hemoglobin, carbon dioxide, carbaminohemoglobin, b. about 20%; **5.** a. bicarbonate, b. about 70%

### Sections 10.5, 10.6

**1.**c; **2.**b; **3.**d; **4.**a; **5.** a. hyperactive immune system causing bronchial muscle contraction and swelling, b. shortness of breath, chest tightness; **6.** a. damage to the alveoli resulting in reduced elasticity, due to either smoking or inheritance, b. difficulty breathing; **7.** a. inflammation of the bronchi, may be due to bacterial infection, b. breathlessness and coughing; **8.** a. lung inflammation due to bacterial or viral infection, b. fever, chills, shortness of breath, cough, pain; **9.** a. bacterial infection beginning in the lungs, may spread to the bloodstream, b. coughing, chest pain, shortness of breath, fever, weight loss; **10.** a. consumption of food contaminated by the bacterium *Clostridium botulinum*, b. difficulty swallowing, vomiting, paralysis of respiratory muscles; **11.** a. usually caused by smoking, may also result from exposure to radon gas or asbestos, b. chest pain, bloody cough; **12.** a. genetic defect resulting in production of thick sticky mucus in the lungs, b. difficulty breathing, respiratory infections

## Chapter Test

**1.**c; **2.**c; **3.**b; **4.**b; **5.**c; **6.**d; **7.**a; **8.**a; **9.**c; **10.**c; **11.**a; **12.**a; **13.**c; **14.**b; **15.**b; **16.**d; **17.**b; **18.**d; **19.**d; **20.**d; **21.**a; **22.**c; **22.**c; **23.**c; **24.**a; **25.**c

# 11

# The Nervous System—Integration and Control

## Chapter Summary and Key Concepts

*After reading and studying this chapter you should know the following:*

**Sections 11.1, 11.2, 11.3, 11.4, 11.5**

1. The three characteristics of the nervous system are (1) it requires information, (2) it integrates information, and (3) it can respond to information very quickly.

2. The two principal parts of the nervous system are the central nervous system (CNS), consisting of the brain and spinal cord, and the peripheral nervous system (PNS), consisting of all nervous tissue not in the CNS.

3. Neurons are nervous system cells specialized to receive and transmit information.

4. The three classifications of neurons are (1) sensory neurons that carry information to the CNS, (2) motor neurons that carry information away from the CNS, and (3) interneurons that communicate between nervous system components.

5. The parts of a neuron are the cell body, dendrites, axon hillock, axon, myelin sheath, nodes of Ranvier, axon terminals, and axon bulbs.

6. The slight difference between the voltage inside and outside a cell is the resting membrane potential.

7. A graded potential occurs when chemically sensitive channels allow ion movements that depolarize or hyperpolarize the membrane.

8. An action potential is a reversal in the voltage difference across a membrane and is triggered when the sum of all graded potentials reaches a threshold level.

9. The three events of an action potential are depolarization, repolarization, and reestablishment of the resting potential.

10. During an action potential, depolarization occurs when sodium moves into the cell, repolarization occurs when potassium moves out of the cell, and reestablishment of the resting membrane potential is accomplished by the sodium-potassium pump.

11. Neurotransmitters are chemicals that carry information across a synapse from one neuron to a target cell. Neurotransmitters that are excitatory depolarize the postsynaptic cell, and neurotransmitters that are inhibitory hyperpolarize the postsynaptic cell.

12. Postsynaptic cells integrate and process information.

**Sections 11.6, 11.7, 11.8, 11.9, 11.10, 11.11, 11.12, 11.13**

13. A nerve contains many axons bundled together. Cranial nerves attach to the brain, and spinal nerves attach to the spinal cord.

14. The PNS is divided into the sensory nervous system that carries information to the CNS, and the motor nervous system that carries information away from the CNS.

15. The motor nervous system branches into the somatic division that controls skeletal muscle, and the autonomic division that controls smooth muscle, cardiac muscle, and glands. The autonomic division branches further into the sympathetic and parasympathetic nervous systems.

16. The three regions of the brain are the hindbrain that includes the medulla oblongata, pons, and cerebellum; the forebrain that includes the thalamus, hypothalamus, limbic system, and cerebrum; and the midbrain.

17. The cerebrum is responsible for higher functions and is divided into the occipital, temporal, parietal, and frontal lobes.

18. The reticular activating system is responsible for levels of sleep and wakefulness. The limbic system in the forebrain is the site of emotion and basic behaviors.

19. The limbic system manages short-term memories for a few hours. Long-term memories occur in the cerebrum and may last for years.

20. Psychoactive drugs affect higher brain functions and can be addictive. Disorders of the nervous system include trauma, infections, abnormal neural activity, and tumors.

# Exercises

*Complete the exercises for each section after you have read and studied the section. If you cannot answer some questions, or answer them incorrectly, return to the chapter and review this information. You may find it helpful to work on only one section at a time. When you have completed all sections, take the Chapter Test as an indicator of your mastery of this topic.*

Chapter 11 *The Nervous System—Integration and Control* 155

**11.1** The nervous system has two principal parts

**11.2** Neurons are the communication cells of the nervous system

**11.3** Neurons initiate action potentials

**11.4** Neuroglial cells support and protect neurons

**11.5** Information is transferred from a neuron to its target

## Matching

\_\_\_\_ 1. **central nervous system**      a. provides support and protection to the neuron

\_\_\_\_ 2. **peripheral nervous system**      b. cells that transmit information between components of the central nervous system

\_\_\_\_ 3. **neurons**      c. the part of a neuron that conducts electrical impulses to facilitate transmission of a message

\_\_\_\_ 4. **sensory neurons**      d. cells of the nervous system specialized for communication

\_\_\_\_ 5. **interneurons**      e. a membrane voltage that triggers an action potential

\_\_\_\_ 6. **motor neurons**      f. the part of a neuron that contains the nucleus

\_\_\_\_ 7. **cell body**      g. movement of an action potential down the axon with transmission occurring between nodes of Ranvier

\_\_\_\_ 8. **resting membrane potential**      h. the part of a neuron that receives incoming messages

\_\_\_\_ 9. **graded potential**      i. an insulating layer that surrounds the axon

\_\_\_\_ 10. **summation**      j. a division of the nervous system consisting of the brain and spinal cord

\_\_\_\_ 11. **threshold**      k. a chemical substance that transmits a signal from a neuron to its target

\_\_\_\_ 12. **action potential**      l. a transient local change in the resting potential of a cell

\_\_\_\_ 13. **synapse**      m. the division of the nervous system containing all components outside the brain and spinal cord

\_\_\_\_ 14. **neurotransmitter**      n. a slight difference in voltage that exists between the inside and the outside of a cell

\_\_\_\_ 15. **dendrites**      o. a sudden, temporary reversal of voltage differences across the cell membrane

\_\_\_\_ 16. **axon**      p. a specialized junction between two cells

\_\_\_\_ 17. **neuroglial cells**      q. neurons that carry information from the CNS to the body

\_\_\_\_ 18. **myelin sheath**      r. the ability of a neuron to respond to the total effect of many incoming signals

\_\_\_\_ 19. **salutatory conduction**      s. neurons that respond to a stimulus and transmit information about the stimulus to the CNS

## Fill-in-the-Blank

*Referenced sections are in parentheses*

20. The _____ pump is important in maintaining the resting membrane potential of cells. (11.3)

21. The resting membrane potential of a neuron is about _____, and the inside of the cell is _____ compared to the outside. (11.3)

22. An action potential occurs because the axon contains _____ sensitive ion channels. (11.3)

23. All action potentials are equal in form and magnitude; this is referred to as the _____ phenomenon. (11.3)

24. The myelin sheaths of neurons in the PNS function to save _____, speed _____, and repair damaged _____. (11.4)

25. Neurons of the CNS are unable to repair injured axons because the myelin sheath is produced by _____ and degenerates when injured. (11.4)

26. The _____ _____ is the cell membrane of the neuron that is sending information at a synapse. (11.5)

27. The _____ _____ is the cell membrane of the neuron that receives information at a synapse. (11.5)

28. Neurotransmitters that hyperpolarize a postsynaptic cell are _____. (11.5)

29. In _____ one neuron transmits a message to many neurons. (11.5)

## Labeling

*Label the indicated areas of Figure 11.1 with a structural term and a function.*

**Structure**

a. axon
b. dendrite
c. cell body
d. axon terminal
e. axon bulb
f. myelin sheath
g. Schwann cell
h. node of Ranvier

**Function**

i. produces myelin
j. transmits information in the form of an electrical impulse
k. contains neurotransmitters in membrane-bound vesicles
l. a terminal branch of the axon
m. receives incoming information
n. a short uninsulated gap in the axon
o. the main body of the neuron
p. an insulating layer surrounding the axon that speeds the rate of conduction

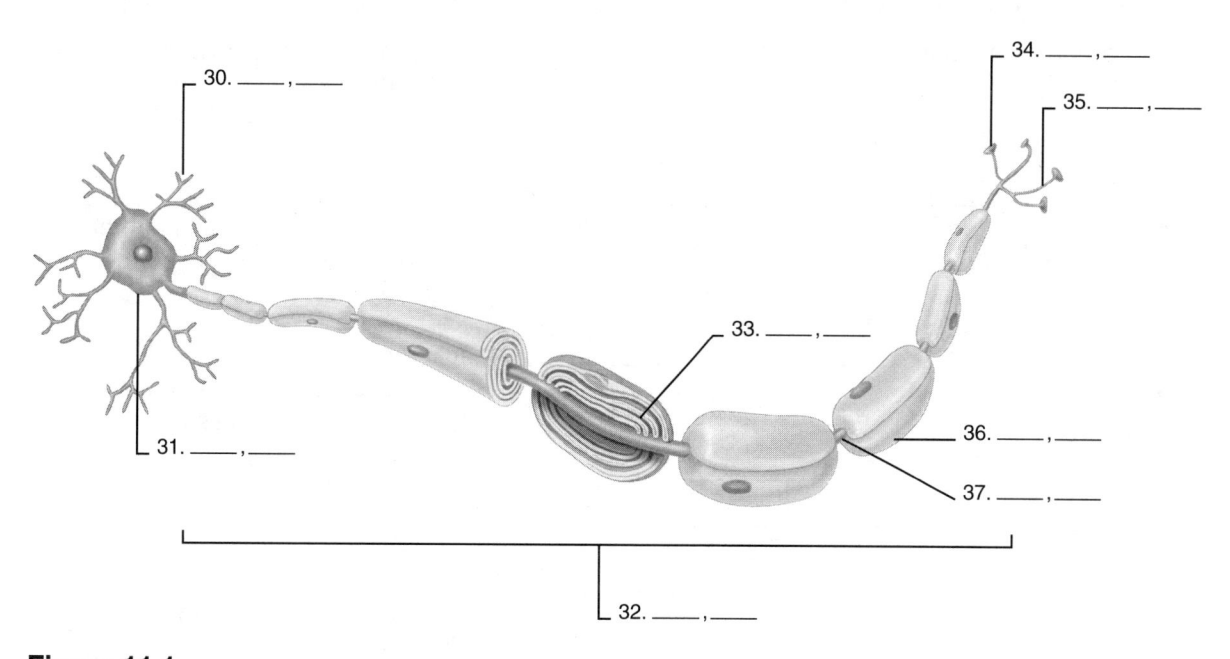

30. ___ , ___
31. ___ , ___
32. ___ , ___
33. ___ , ___
34. ___ , ___
35. ___ , ___
36. ___ , ___
37. ___ , ___

**Figure 11.1**

## Completion

*Complete the table below as you refer to Figure 11.2. The first row has been completed as an example.*

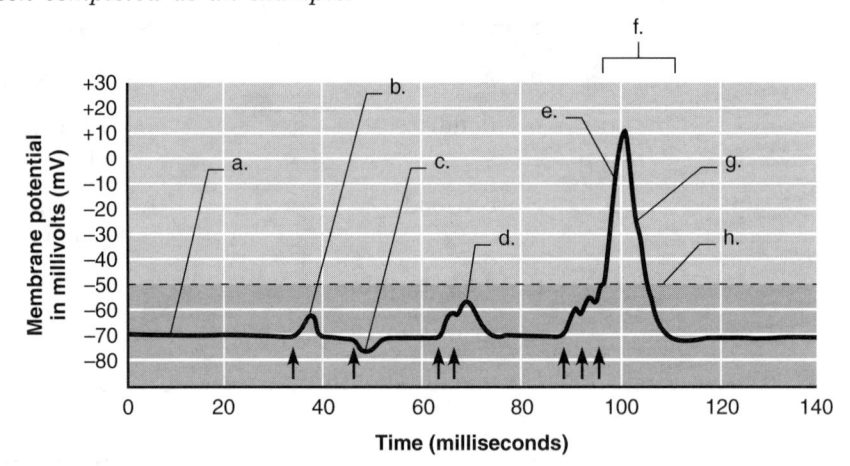

**Figure 11.2**

| Name of Event | Region of Figure 11.2 | Description of Electrical Activity |
|---|---|---|
| 38. resting membrane potential | 39. a | a difference in voltage between the inside and outside of a cell, usually about −70 mV |
| 40. | 41. | a graded potential that makes the membrane voltage more negative |
| 42. | 43. | potassium ions move out of the axon as potassium channels open |
| 44. | 45. | electrical event in the axon—a reversal of the voltage difference across the membrane |
| 46. | 47. | voltage level achieved by summation that triggers an action potential |

| Name of Event | Region of Figure 11.2 | Description of Electrical Activity |
|---|---|---|
| 48. | 49. | a graded potential that makes the membrane voltage less negative |
| 50. | 51. | sodium ions move into the axon as sodium channels open |
| 52. | 53. | the cumulative effect of many graded potentials |

## Word Choice

*Circle the term or phrase that correctly completes each sentence.*

54. Graded potentials (are/are not) self-propagating.

55. The opening of sodium channels on the axon allows $Na^+$ to move (into/out of) the cell.

56. An action potential (is/is not) self-propagating.

57. The speed of an action potential increases as the diameter of the axon (increases/decreases).

58. When an action potential arrives at the axon bulb, calcium channels (open/close) and $Ca^{++}$ moves (into/out of) the bulb.

59. The effect of a neurotransmitter is relatively (short-lived/long-lived).

60. Muscle cells (do/do not) process and integrate information because they receive input from (one/multiple) neuron(s).

61. Excitatory neurotransmitters (depolarize/repolarize) the (presynaptic/postsynaptic) cell.

11.6 The PNS relays information between tissues and the CNS

11.7 The brain and spinal cord comprise the CNS

## Crossword Puzzle

**Across**

1. The cell bodies of _____ neurons lie outside the CNS
2. The cell bodies of _____ neurons lie in the CNS
3. The branch of the autonomic nervous system that prepares the body for "fight or flight"
7. Connective tissue membranes that surround the CNS
10. Cerebrospinal fluid is secreted into the _____ of the brain
12. Bundles of many axons wrapped in a connective sheath
13. _____ nerves that connect directly with the brain
14. The root of a spinal nerve that contains sensory neurons
16. A spinal _____ is an involuntary response that bypasses the brain
17. _____ nerves connect with the spinal cord

**Down**

1. The branch of the automatic nervous system that predominates during relaxation
4. The branch of the PNS that controls smooth and cardiac muscle
5. Pathways of the autonomic nervous system require _____ neurons
6. The branch of the PNS that controls skeletal muscle
8. The _____ matter of the CNS contains nerve tracts
9. The root of a spinal nerve that contains motor neurons
11. A collection of cell bodies of postsynaptic neurons
15. The _____ matter of the CNS contains unmyelinated axons and cell bodies

11.8 **The brain processes and acts on information**

11.9 **Brain activity continues during sleep**

11.10 **The limbic system is the site of emotion and basic behaviors**

11.11 **Memory involves storing and retrieving information**

## Fill-in-the-Blank

*All statements reference section 11.8.*

1. The three major functional divisions of the brain are the _____, _____, and _____.

2. The oldest and most primitive division of the brain is the _____.

3. The three components of the hindbrain are the _____ _____, _____, and _____.

4. The cardiovascular center is located in the _____ _____.

5. Motor nerves that leave the forebrain cross over to the other side of the body in the _____ _____.

6. Wakefulness is maintained by the _____ _____.

7. Four important areas of the brain located in the forebrain include the _____, _____, _____ _____, and _____.

8. The _____ helps to regulate homeostasis by receiving sensory information from the body.

9. Short-term memory is created in the _____ _____.

10. The _____ lobe of the brain initiates muscle activity and conscious thought.

## Labeling

*Label the indicated areas in Figure 11.3 with a structural term and a function.*

**Structure**

a. pons
b. cerebrum
c. cerebellum
d. thalamus
e. medulla oblongata
f. corpus callosum

**Function**

g. controls automatic functions of the internal organs
h. receives, processes, and transfers information
i. aids the medulla in regulating respiration
j. bridges the two cerebral hemispheres
k. involved in decision making and conscious thought
l. coordinates basic subconscious body movements

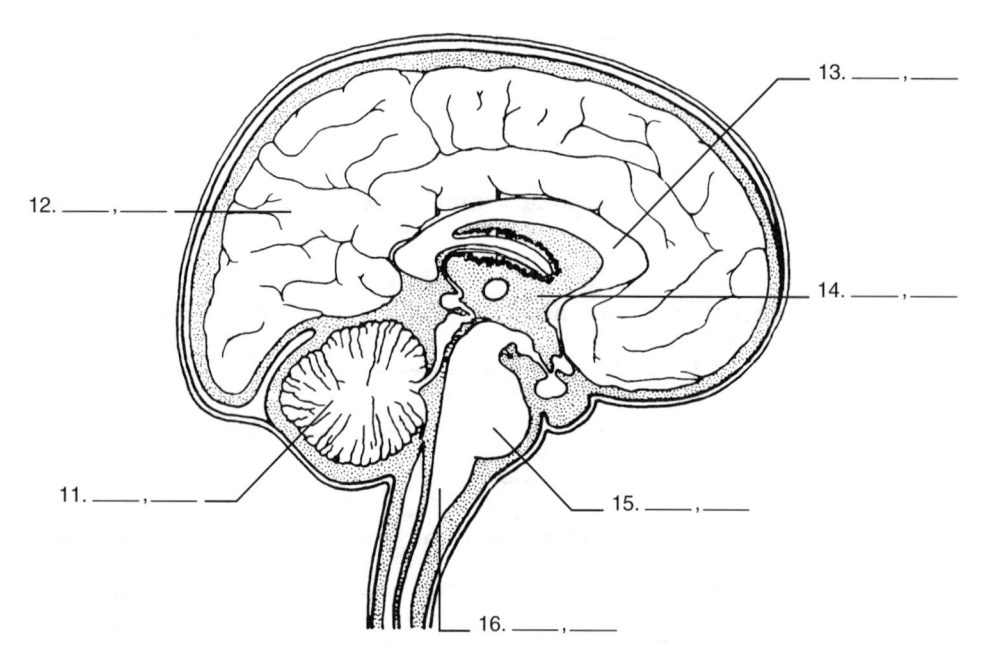

12. ___ , ___
13. ___ , ___
14. ___ , ___
11. ___ , ___
15. ___ , ___
16. ___ , ___

**Figure 11.3**

### 11.12 Psychoactive drugs affect higher brain functions
### 11.13 Disorders of the nervous system

## Short Answer

1. What is a psychoactive drug?

2. Why can psychoactive drugs affect the brain when many other drugs cannot?

3. What is the common mode of action for psychoactive drugs?

4. What is the affect of cocaine on the brain?

5. Why does tolerance develop?

6. Define "addiction."

## Completion

*Name each disorder of the nervous system described in the table.*

| Disorder | Description |
| --- | --- |
| 7. | blow to the head or neck with loss of consciousness |
| 8. | violent trauma to the vertebrae that compresses, tears, or severs the spinal cord |
| 9. | an infection resulting in inflammation of the brain |
| 10. | an infection resulting in inflammation of the meninges |

| Disorder | Description |
| --- | --- |
| 11. | viral brain infection transmitted to humans by direct contact with infected animals, often through a bite |
| 12. | recurrent episodes of abnormal electrical activity in the brain, often of unknown causes |
| 13. | progressive degenerative disorder caused by the loss of dopamine-releasing neurons |
| 14. | changes in structure and activity of frontal and temporal lobes leading to mental impairment |
| 15. | abnormal growth in or on the brain |

# Chapter Test

## Multiple Choice

1. Which of the following does not apply to the peripheral nervous system?.
   a. Components are found outside the brain and spinal cord
   b. Includes a sensory and motor division.
   c. The somatic division controls smooth muscle activity.
   d. The sympathetic and parasympathetic division work antagonistically.

2. Neurons of the PNS that transmit information to the CNS are:
   a. sensory neurons.
   b. motor neurons.
   c. interneurons.
   d. sympathetic neurons.

3. The _____ of a neuron carry information toward the cell body, while the _____ carries information away from the cell body.
   a. myelin sheath, synaptic knob
   b. dendrites, axon
   c. axon, dendrites
   d. postsynaptic membrane, presynaptic membrane

4. The myelin sheath of PNS neurons is produced by _____, while the myelin sheath of CNS neurons is produced by _____.
   a. oligodendrocytes, Schwann cells
   b. neuroglia, oligodendrocytes
   c. Schwann cells, oligodendrocytes
   d. Schwann cells, Schwann cells

5. The sodium-potassium pump:
   a. transports 3 $Na^+$ out of the cell and 2 $K^+$ into the cell.
   b. transports 3 $Na^+$ into the cell and 2 $K^+$ out of the cell.
   c. helps to maintain the resting membrane potential of a cell.
   d. a and c
   e. b and c

6. Which of the following characteristics of the neuron cell membrane contribute to the maintenance of the resting membrane potential?
   a. The membrane is more permeable to sodium than to potassium.
   b. Sodium leaks out of the membrane more easily than potassium.
   c. The sodium-potassium pump transports three sodium ions out of the cell for every two potassium ions it pumps into the cell.
   d. a and c
   e. all of the above

7. Graded potentials:
   a. are permanent changes in the resting potential of a membrane.
   b. occur in a local area of the membrane.
   c. always depolarize the membrane.
   d. grow stronger as they spread through the cell.

8. Summation may occur:
   a. when rapid stimulation of a neuron causes the membrane voltage to reach threshold levels.
   b. when many neurons all cause hyperpolarizing graded potentials.
   c. when the presence of the axon hillock lowers the normal threshold level.
   d. when the cell body contains both chemically sensitive and voltage-sensitive channels.

9. Repolarization occurs when:
   a. voltage-sensitive calcium channels open.
   b. chemically sensitive sodium channels close.
   c. potassium channels open.
   d. sodium leaks out of the cell.

10. Which of the following would occur in a neuron if calcium could not enter the axon bulb?
    a. absence of graded potentials
    b. absence of action potentials
    c. inability to manufacture neurotransmitters
    d. inability to release neurotransmitters

11. The absolute refractory period:
    a. occurs during transmission of an action potential.
    b. occurs shortly after transmission of an action potential.
    c. occurs shortly before transmission of an action potential.
    d. reflects the actions of the sodium-potassium pump.

12. Action potentials are "all-or-none" events. Which of the following statements contradicts this?
    a. Besides generating an action potential, the only additional information a neuron can transmit is the frequency of stimulation.
    b. If a threshold level of stimulation is reached, the action potential will always be of the same form and strength.
    c. Summation in a postsynaptic neuron can occur when the frequency of stimulation increases.
    d. The level of neurotransmitter bound to the postsynaptic membrane affects the speed of transmission of the action potential.

13. Excitatory neurotransmitters:
    a. depolarize the postsynaptic membrane.
    b. hyperpolarize the postsynaptic membrane.
    c. always initiate an action potential.
    d. can depolarize or hyperpolarize the postsynaptic membrane depending on the type of receptors present.

14. Which of the following must occur immediately before neurotransmitter can be released into the synaptic cleft?
    a. an action potential
    b. a graded potential
    c. reuptake of neurotransmitter by the presynaptic membrane
    d. opening of calcium channels on the presynaptic membrane

15. When one neuron receives input from many neurons, this is called:
    a. summation.
    b. convergence.
    c. divergence.
    d. multiple stimulation.

16. The ventral root of a spinal nerve contains:
    a. sensory neurons.
    b. motor neurons.
    c. sensory and motor neurons.
    d. cell bodies only.

17. Motor pathways requiring a preganglionic and a postganglionic neuron belong to the _____ nervous system.
    a. somatic division of the
    b. sensory division of the
    c. central
    d. autonomic

18. If your heart rate and blood pressure increase in response to a crisis, you are experiencing the immediate effects of the _____ nervous system.
    a. sensory
    b. sympathetic
    c. parasympathetic
    d. reflexive

19. Which of the following is *not* one of the meninges of the central nervous system?
    a. dura mater
    b. pia mater
    c. ventral
    d. arachnoid

20. White matter of the spinal cord contains _____,
    while gray matter contains _____.
    a. nerve tracts, cell bodies
    b. cell bodies, nerves
    c. unmyelinated axons, myelinated axons
    d. motor neurons, sensory neurons

21. Which of the following major functional divisions of the brain are incorrectly matched with its function?
    a. midbrain—coordinates muscle groups
    b. hindbrain—responds to light and sound
    c. forebrain—integrates sensory information
    d. all of the above are correctly matched

22. Long term memory occurs when:
    a. neurons are permanently changed.
    b. the memory is stored in the cerebellum.
    c. dopamine remains in the synaptic cleft.
    d. convergence strengthens the excitatory effect of a neurotransmitter.

23. In the primary somatosensory area:
    a. the brain responds to visual input.
    b. conscious thought is coordinated with motor responses.
    c. body parts that are extremely sensitive are represented by a greater area than less sensitive body parts.
    d. body parts that are coordinated with motor responses are represented by a greater area than body parts that are not.

24. During which stage of sleep do skeletal muscles relax and eye and body movements cease?
    a. stage 1
    b. stage 2
    c. stage 3
    d. stage 4
    e. stage 5

25. The site of emotions and basic behavior is the:
    a. parietal lobe of the cerebral cortex.
    b. thalamus.
    c. limbic system.
    d. hindbrain.

# Key Concept Review Questions

*Each of the Key Concepts listed at the beginning of this chapter has been rewritten as a question below. After successfully completing the study guide exercises and the Chapter Test, you should be able to answer each of these questions. Refer to the Key Concepts list at the beginning of this chapter to check your answers.*

1. What are the three characteristics of the nervous system?
2. What are the two principal parts of the nervous system and what components does each contain?
3. What cells of the nervous system are specialized to receive and transmit information?
4. What are the three classifications of neurons?
5. What are the parts of a neuron?
6. What is the resting membrane potential?
7. What causes a graded potential to occur?
8. What is an action potential?
9. What are the three events of an action potential?
10. During an action potential, what ion activity causes each of the three events?
11. What is a neurotransmitter? How do excitatory neurotransmitters differ from inhibitory neurotransmitters in their effects on the postsynaptic cell membrane?
12. What synaptic cells integrate and process information?
13. What is a nerve? How do spinal nerves differ from cranial nerves?
14. What are the two divisions of the peripheral nervous system? Which division carries information to the brain? Which division carries information away from the brain?
15. What are the branches of the motor nervous system? What are the branches of the autonomic division?
16. What are the regions of the brain?
17. What is the general function of the cerebrum? List the lobes of the cerebrum.
18. What is the function of the reticular activating system? What is the function of the limbic system?
19. How does short-term memory differ from long-term memory in location and duration?
20. How to psychoactive drugs affect the brain?
21. What are the four general categories of nervous system disorders discussed in this chapter?

# Answer Key

## Sections 11.1, 11.2, 11.3, 11.4, 11.5

**1.** j; **2.** m; **3.** d; **4.** s; **5.** b; **6.** q; **7.** f; **8.** n; **9.** l; **10.** r; **11.** e; **12.** o; **13.** p; **14.** k; **15.** h; **16.** c; **17.** a; **18.** i; **19.** g; **20.** sodium-potassium; **21.** −70mV, negative; **22.** voltage-; **23.** all-or-none; **24.** energy, transmission, axons; **25.** oligodendrocytes; **26.** presynaptic membrane; **27.** postsynaptic membrane; **28.** inhibitory; **29.** divergence; **30.** b,m; **31.** c,o; **32.** a,j; **33.** g,I; **34.** e,k; **35.** d,l; **36.** f,p; **37.** h,n; **38.** given as an example; **39.** a; **40.** hyperpolarizing graded potential; **41.** c; **42.** repolarization phase of an action potential; **43.** g; **44.** action potential; **45.** f; **46.** threshold; **47.** h; **48.** depolarizing graded potential; **49.** b; **50.** depolarization phase of an action potential; **51.** e; **52.** summation; **53.** d; **54.** are not; **55.** into; **56.** is; **57.** increases; **58.** open, into; **59.** short-lived; **60.** do not, one; **61.** depolarize, postsynaptic

## Sections 11.6, 11.7

**Crossword Puzzle: 1.** (across) postganglionic; **1.** (down) parasympathetic; **2.** preganglionic; **3.** sympathetic; **4.** autonomic; **5.** two; **6.** somatic; **7.** meninges; **8.** white; **9.** ventral; **10.** ventricle; **11.** ganglia; **12.** nerves; **13.** Cranial; **14.** dorsal; **15.** gray; **16.** reflex; **17.** Spinal

## Sections 11.8, 11.9, 11.10, 11.11

**1.** hindbrain, midbrain, forebrain; **2.** hindbrain; **3.** medulla oblongata, cerebellum, pons; **4.** medulla oblongata; **5.** medulla oblongata; **6.** reticular formation; **7.** hypothalamus, thalamus, limbic system, cerebrum; **8.** hypothalamus; **9.** limbic system; **10.** frontal; **11.** c,l; **12.** b,k; **13.** f,j; **14.** d,h; **15.** a,i; **16.** e,g

## Sections 11.12, 11.13

**1.** A drug that affects higher brain functions; **2.** they cross the blood-brain barrier; **3.** they influence the concentration or mode of action of brain neurotransmitters; **4.** cocaine prolongs the presence of dopamine in the synaptic cleft, causing repeated stimulation of a neuron; **5.** the liver detoxifies drugs; rising levels of liver enzymes necessitate higher dosages to achieve the same effect; **6.** the need to continue obtaining and using a substance despite one's better judgment and good intentions; **7.** concussion; **8.** spinal cord injury; **9.** encephalitis; **10.** meningitis; **11.** rabies; **12.** epilepsy; **13.** Parkinson's disease; **14.** Alzheimer's disease; **15.** brain tumor

## Chapter Test

**1.** c; **2.** a; **3.** b; **4.** c; **5.** d; **6.** c; **7.** b; **8.** a; **9.** c; **10.** d; **11.** a; **12.** d; **13.** a; **14.** d; **15.** b; **16.** b; **17.** d; **18.** b; **19.** c; **20.** b; **21.** d; **22.** a; **23.** c; **24.** b; **25.** c

# 12

# Sensory Mechanisms

## Chapter Summary and Key Concepts

*After reading and studying this chapter you should know the following:*

### Sections 12.1, 12.2, 12.3

1. Receptors convert the energy of a stimulus into impulses in sensory neurons.

2. Receptors are classified according to the type of stimulus energy they convert.

3. The brain interprets nerve impulses based on their origin and frequency.

4. Adaptation occurs in some receptors and is the cessation of impulses in the presence of a continued stimulus.

5. The somatic senses are temperature, touch, vibration, pressure, pain, and awareness of body position and movements. The somatic senses arise from receptors located throughout the body.

6. The special senses are taste, smell, hearing, balance, and vision. The special senses arise from receptors in specialized areas of the body.

7. Taste depends on chemoreceptors located in taste buds on the tongue and mouth.

8. Smell depends on chemoreceptors located in the nasal passages.

### Sections 12.4, 12.5

9. Mechanoreceptors in the ear detect sound waves.

10. The ear consists of an outer ear, middle ear, and inner ear.

11. The outer ear is called the pinna and channels sound waves.

12. The middle ear consists of three small bones that amplify sound and transmit the waves to the inner ear.

13. The inner ear converts sound to impulses in the cochlea and transmits information to the brain via the auditory nerve.

14. The inner ear also contains the semicircular canals that sense rotational movement, and the vestibular apparatus that senses static head position and linear acceleration.

**Sections 12.6, 12.7**

15. The structures of the eye convert visual stimuli to impulses that are transmitted to the brain via the optic nerve.
16. Light entering the eye is bent by the cornea and the lens to focus an image.
17. Accommodation is the adjustment of lens curvature that allows the eye to focus on objects at any distance.
18. The photoreceptor cells of the eye are the photopigment-containing rods and cones.
19. Rods function in dim light, and cones respond to color.
20. Disorders of sensory mechanisms include retinal detachment, glaucoma, cataracts, color blindness, otitis media, and Ménière's syndrome.

# Exercises

*Complete the exercises for each section after you have read and studied the section. If you cannot answer some questions, or answer them incorrectly, return to the chapter and review this information again. You may find it helpful to work on only one section at a time. When you have completed all sections, take the Chapter Test as an indicator of your mastery of this topic.*

**12.1  Receptors receive and convert stimuli**

**12.2  Somatic sensations arise from receptors throughout the body**

**12.3  Taste and smell depend on chemoreceptors**

## Matching

_____ 1. **stimulus**           a. receptors that respond to forms of mechanical energy

_____ 2. **receptor**            b. receptors that respond to light

_____ 3. **sensation**           c. sensory neurons located in the nasal passages

_____ 4. **perception**          d. mechanoreceptors that monitor muscle length

_____ 5. **mechanoreceptors**    e. sensory input that causes a change within or outside the body

_____ 6. **thermoreceptors**     f. a cluster of taste cells and support cells on the tongue and mouth

_____ 7. **pain receptors**      g. sensations of temperature, touch, vibration, pressure, and pain

_____ 8. **chemoreceptors**      h. a structure specialized to receive a certain stimulus

_____ 9. **photoreceptors**      i. sensory neurons stop sending impulses while the stimulus is still present

_____ 10. **receptor adaptation**   j. conscious awareness of a stimulus

_____ 11. **somatic sensation**     k. interpretation of a sensation

_____ 12. **special senses**        l. receptors that respond to the presence of chemicals

_____ 13. **muscle spindles**       m. receptors that respond to heat or cold

_____ 14. **taste buds**            n. sensations of taste, smell, hearing, balance, and vision

_____ 15. **olfactory receptor cells**   o. receptors that respond to tissue damage or excessive pressure or temperature

## Word Choice

*Circle the term or phrase that correctly completes each sentence.*

16. Individuals are (conscious/not conscious) of receptors that function in negative feedback loops to maintain homeostasis.

17. Receptors in the skin for light touch and pressure adapt (slowly/quickly).

18. Body parts with the greatest sensory sensitivity involve the (fewest/most) neurons.

19. Receptors for (slow pain/fast pain) are usually located near the surface of the body.

## Fill-in-the-Blank

*Referenced sections are in parentheses.*

20. Taste buds convert a chemical signal into a(n) _____ _____. (12.3)

21. The four major categories of taste are _____, _____, _____, and _____. (12.3)

22. Taste buds contain chemoreceptors for _____, while olfactory receptor cells have chemoreceptors for _____. (12.3)

23. The receptors of olfactory cells are modified dendritic endings of sensory neurons and are called _____ _____. (12.3)

174  Study Guide for *Human Biology*

**12.4 Hearing: Mechanoreceptors detect sound waves**

**12.5 The inner ear plays an essential role in balance**

## Labeling

*Use the terms to label Figure 12.1.*

| middle ear | pinna | malleus |
| auditory canal | semicircular canals | oval window |
| outer ear | cochlear nerve | vestibular nerve |
| incus | cochlea | tympanic membrane |
| vestibule | inner ear | stapes |
| round window | eustachian tube | |

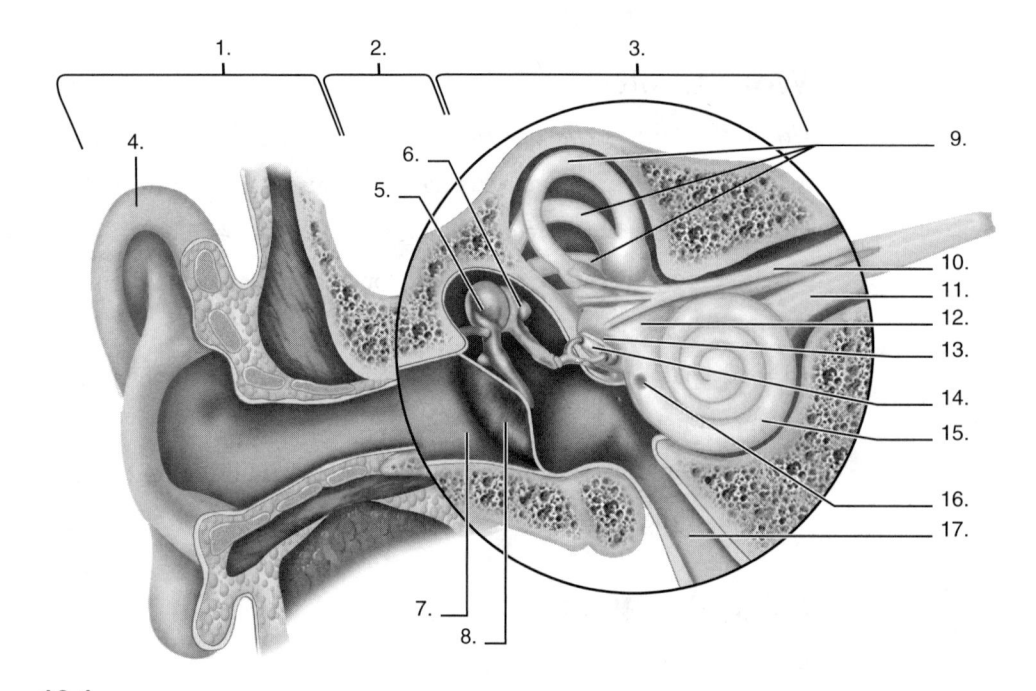

**Figure 12.1**

1. _____
2. _____
3. _____
4. _____
5. _____
6. _____
7. _____
8. _____
9. _____

10. _____
11. _____
12. _____
13. _____
14. _____
15. _____
16. _____
17. _____

## Paragraph Completion

*Use the terms to complete the paragraph, and then repeat this exercise with the terms covered.*

| | | |
|---|---|---|
| amplitude | eardrum | neurotransmitter |
| pinna | malleus | vestibular |
| basilar | stapes | cochlea |
| waves | oval window | auditory |
| tectorial | incus | frequency |
| base | round window | tip |

Sounds are (18) _____ of compressed air. The loudness of a sound depends on the (19) _____ of the waves; the tone of a sound depends on the (20) _____ of the waves. Sound waves first arrive at the (21) _____ and cause vibrations of the (22) _____. These vibrations are passed on to the (23) _____, (24) _____, and (25) _____ of the middle ear, which amplify the sound and lead to vibration of a small membrane called the (26) _____. This membrane transmits the vibrations to the fluid in the (27) _____ canal of the (28) _____. Pressure waves in the fluid of the cochlea vibrate the (29) _____ membrane, moving the hair cells that are embedded in the (30) _____ membrane. Bending of the hair cells causes the release of a (31) _____ and generates an impulse in the (32) _____ nerve. High-pitched tones vibrate the basilar membrane closer to the (33) _____ of the cochlea, while low-pitched tones vibrate the membrane closer to the (34) _____ of the cochlea. Fluid vibrations are dissipated when they reach the (35) _____.

## Completion

Write the letters next to the structures in the following list beside the correct region and function in the table below. Some structures may be used more than once.

- a. cochlea
- b. cupula
- c. hair cells
- d. malleus
- e. stapes
- f. saccule
- g. incus
- h. ampulla
- i. eardrum
- j. auditory tube
- k. auditory canal
- l. oval window
- m. cochlear duct
- n. pinna
- o. utricle
- p. vestibule
- q. utricle
- r. organ of Corti
- s. round window
- t. tectorial membrane
- u. semicircular canals
- v. basilar membrane
- w. otoliths
- x. vestibular canal
- y. tympanic canal

| Region of the Ear | Functions | Structures |
| --- | --- | --- |
| Outer ear | Receives and channels sound | 36. |
| Middle ear | Amplifies sound | 37. |
| Inner ear | Sorts sounds by tone and converts them into impulses | 38. |
| Inner ear | Senses rotational movements and converts them into impulses | 39. |
| Inner ear | Sense static position and linear movement and converts them into impulses | 40. |

**12.6 Vision: Detecting and interpreting visual stimuli**

**12.7 Disorders of sensory mechanisms**

## Labeling

Use the terms to label Figure 12.2.

| pupil | choroid | iris | aqueous humor |
| sclera | fovea | retina | Canal of Schlemm |
| optic disk | lens | cornea | ciliary muscle |
| optic nerve | vitreous humor | | |

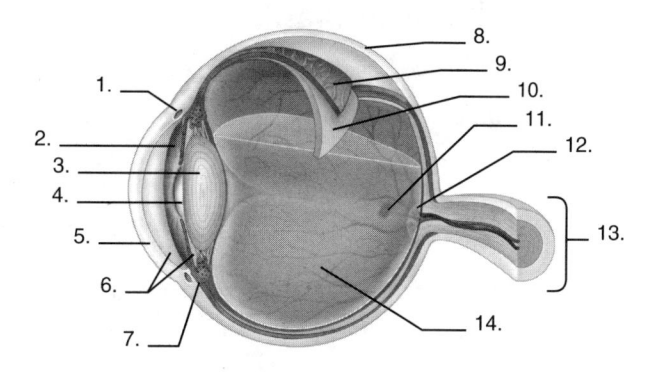

**Figure 12.2**

1. _____
2. _____
3. _____
4. _____
5. _____
6. _____
7. _____
8. _____
9. _____
10. _____
11. _____
12. _____
13. _____
14. _____

## Matching

*Match each of the following structures of the eye with the correct function.*

____ 15. **sclera**
____ 16. **cornea**
____ 17. **aqueous humor**
____ 18. **iris**
____ 19. **lens**
____ 20. **ciliary muscle**
____ 21. **vitreous humor**
____ 22. **retina**
____ 23. **retinal rods**
____ 24. **retinal cones**
____ 25. **fovea**
____ 26. **optic disk**
____ 27. **optic nerve**
____ 28. **choroids**

a. absorbs light and converts it into impulses
b. contains greatest concentration of photoreceptors
c. nourishes and cushions cornea and lens
d. regulates focus
e. covers and protects eyeball
f. "blind spot" where optic nerve exits eye
g. transmits light to retina
h. nourishes retina and absorbs light not absorbed by retinal photoreceptors
i. photoreceptors responsible for black-and-white vision in dim light
j. transmits impulses to the brain
k. bends incoming light and focuses light
l. photoreceptors responsible for color vision and high visual acuity
m. adjusts amount of incoming light
n. adjusts curvature of the lens

## Fill-in-the-Blank

*Referenced sections are in parentheses.*

29. The flattened disks of photoreceptor cells contain a protein called a _____. (12.6)

30. _____ is the photopigment found in rods. (12.6)

31. _____ are photoreceptor cells that respond to color. (12.6)

32. Light entering the eye is bent first by the _____ and then by the _____. (12.6)

33. The innermost layer of the retina consists of neurons called _____ _____. (12.6)

34. Nearsightedness, also called _____, occurs when the eyeball is too _____. (12.6)

## Completion

Write the name of each sensory disorder described in the table.

| Sensory Disorder | Description |
| --- | --- |
| 35. | the retina separates from the choroid, often caused by a blow to the head |
| 36. | blockage in the Canal of Schlemm prevents drainage of the aqueous humor, leading to pressure in the eye |
| 37. | a decrease in the normal transparency of the lens |
| 38. | inflammation of the middle ear |
| 39. | a chronic condition of the inner ear, characterized by recurrent episodes of dizziness, nausea, and hearing loss |
| 40. | distant objects are focused in front of the retina |

# Chapter Test

## Multiple Choice

1. Receptors are classified according to:
   a. their location in the body.
   b. the type of cell they arise from.
   c. the type of stimulus energy they convert.
   d. how easily they generate an action potential.

2. Pain receptors adapt slowly or not at all because:
   a. lack of adaptation is important to survival.
   b. only mechanoreceptors are capable of adaptation.
   c. adaptation occurs but is ignored by the brain.
   d. the stimulus is too strong for adaptation to occur.

3. Meissner corpuscles detect:
   a. light touch.
   b. deep pressure.
   c. temperature.
   d. ongoing pressure.

4. Muscle spindles are:
   a. chemoreceptors.
   b. photoreceptors.
   c. thermoreceptors.
   d. mechanoreceptors.

5. Which of the following is not a somatic sensation?
   a. balance
   b. vibration
   c. pain
   d. temperature

6. Chemoreceptors for taste are specific for a chemical called a:
   a. tastant.
   b. papillae.
   c. taste bud.
   d. taste hair.

7. The chemoreceptors for smell are located on:
   a. sensory neurons.
   b. epithelial cells.
   c. naked nerve endings.
   d. glandular cells.

8. The olfactory bulb is:
   a. the site of origin for olfactory receptor cells.
   b. an area of the nose where olfactory cells are highly concentrated.
   c. an area of the brain where olfactory receptor cells synapse with other neurons.
   d. the area of the brain where smells are interpreted.

9. Sounds are waves of compressed air. The loudness of a sound is determined by:
   a. the frequency of the waves.
   b. the amplitude of the waves.
   c. the temperature of the air.
   d. the volume of air.

10. The pinna and auditory canal are part of the:
    a. inner ear.
    b. middle ear.
    c. outer ear.
    d. temporal lobe.

11. Which part of the ear generates impulses for sound?
    a. ampulla
    b. utricle
    c. saccule
    d. cochlea

12. Vibrations are passed from the middle ear to the inner ear through the:
    a. cochlear window.
    b. round window.
    c. oval window.
    d. tympanic window.

13. The base of the cochlear duct is formed by the:
    a. tectorial membrane.
    b. vestibular membrane.
    c. cochlear membrane.
    d. basilar membrane.

14. The structure of the inner ear that converts pressure waves to impulses is the:
    a. vestibular canal.
    b. cochlear duct.
    c. organ of Corti.
    d. round window.

15. The mechanoreceptor cells of the ear are:
    a. hair cells.
    b. tectorial cells.
    c. vestibular cells.
    d. tympanic cells.

16. The semicircular canals sense:
    a. rotational movement of the head and body.
    b. inertia.
    c. linear acceleration.
    d. vertical deceleration.

17. Which of the following structural layers of the eye is incorrectly matched with its function?
    a. sclera—produces the aqueous humor
    b. retina—contains photoreceptor cells
    c. choroids—contains blood vessels that nourish the retina
    d. all are correctly matched with their function

18. The photoreceptors of the eye are:
    a. the optic nerve.
    b. the ciliary muscle.
    c. rods and cones.
    d. choroid pigment cells.

19. What structure of the eye determines how much light enters the eye?
    a. cornea
    b. pupil
    c. iris
    d. lens

20. The fluid found between the lens and the cornea is the:
    a. aqueous humor.
    b. vitreous humor.
    c. ciliary fluid.
    d. synovial fluid.

21. Accommodation occurs as a result of the actions of the:
    a. cornea.
    b. ciliary muscle.
    c. vitreous humor.
    d. retina.

22. The layer of bipolar cells in the retina:
    a. receives information from the rods and cones.
    b. receives information from ganglion cells.
    c. absorbs light to stimulate rods and cones.
    d. possesses axons that will become the optic nerve.

23. Rhodopsin is:
    a. an enzyme produced by the ciliary muscle.
    b. a nutrient found in the vitreous humor.
    c. the photopigment found in rods.
    d. a chemoreceptor that monitors blood supply to the retina.

24. The optic disk is the area of the eye:
    a. with the highest concentration of cones.
    b. with the highest concentration of rods.
    c. where the optic nerve exits.
    d. where the retina attaches to the sclera.

25. Ménière's syndrome is a condition in which:
    a. the retina detaches from the choroids.
    b. an individual suffers from dizziness and hearing loss.
    c. the middle ear is inflamed.
    d. olfactory receptors are damaged.

# Key Concept Review Questions

*Each of the Key Concepts listed at the beginning of this chapter has been rewritten as a question below. After successfully completing the study guide exercises and the Chapter Test, you should be able to answer each of these questions. Refer to the Key Concepts list at the beginning of this chapter to check your answers.*

1. What do receptors do with a stimulus?
2. How are receptors classified?
3. What characteristics does the brain use to interpret a nerve impulse?
4. What is adaptation? Does it occur in all receptors?
5. What are the somatic senses? Where are the receptors for somatic senses located?
6. What are the special senses? Where are the receptors for the special senses located?
7. What type of receptor is responsible for taste? Where are these receptors located?

8. What type of receptor is responsible for smell? Where are these receptors located?

9. What type of receptors detect sound waves?

10. What are the three main parts of the ear?

11. What is the outer ear called? What is the function of the outer ear?

12. What structures are found in the middle ear? What is the function of the middle ear?

13. What role does the inner ear play in sound? What general part of the inner ear is responsible for hearing?

14. What part of the inner ear functions in sensing rotational movement? What part of the inner ear functions in sensing head position and linear acceleration?

15. What does the eye do with visual stimuli?

16. What structures of the eye bend light to focus an image?

17. What is accommodation?

18. What are the photoreceptor cells of the eye?

19. What is the specific function of rods and cones?

20. What are six disorders of sensory mechanisms discussed in this chapter?

## Answer Key

### Sections 12.1, 12.2, 12.3

**1.** e; **2.** h; **3.** j; **4.** k; **5.** a; **6.** m; **7.** o; **8.** l; **9.** b; **10.** i; **11.** g; **12.** n; **13.** d; **14.** f; **15.** c; **16.** not conscious; **17.** quickly; **18.** most; **19.** fast pain; **20.** action potential; **21.** sweet, sour, salty, bitter; **22.** tastants, odorants; **23.** olfactory hairs

### Sections 12.4, 12.5

**1.** outer ear; **2.** middle ear; **3.** inner ear; **4.** pinna; **5.** malleus; **6.** incus; **7.** auditory canal; **8.** tympanic membrane; **9.** semicircular canals; **10.** vestibular nerve; **11.** cochlear nerve; **12.** vestibule; **13.** oval window; **14.** stapes; **15.** cochlea; **16.** round window; **17.** Eustachian tube; **18.** waves; **19.** amplitude; **20.** frequency; **21.** pinna; **22.** eardrum; **23.** malleus; **24.** incus; **25.** stapes; **26.** oval window; **27.** vestibular; **28.** cochlea; **29.** basilar; **30.** tectorial; **31.** neurotransmitter; **32.** auditory; **33.** base; **34.** tip; **35.** round window; **36.** k,n; **37.** d,e,g,i,j,l; **38.** a,c,m,r,s,t,v,x,y; **39.** b,c,h,u; **40.** c,f,p,q,w

## Sections 12.6, 12.7

**1.** Canal of Schlemm; **2.** iris; **3.** lens; **4.** pupil; **5.** cornea; **6.** aqueous humor; **7.** ciliary muscle; **8.** sclera; **9.** choroids; **10.** retina; **11.** fovea; **12.** optic disk; **13.** optic nerve; **14.** vitreous humor; **15.**e; **16.**k; **17.**c; **18.**m; **19.**d; **20.**n; **21.**g; **22.**a; **23.**i; **24.**l; **25.**b; **26.**f; **27.**j; **28.**h; **29.** photopigment; **30.** Rhodopsin; **31.** Cones; **32.** cornea, lens; **33.** ganglion cells; **34.** myopia, long; **35.** retinal detachment; **36.** glaucoma; **37.** cataracts; **38.** otitis media; **39.** Ménière's syndrome; **40.** myopia

## Chapter Test

**1.**c; **2.**a; **3.**b; **4.**d; **5.**a; **6.**a; **7.**a; **8.**c; **9.**b; **10.**c; **11.**d; **12.**c; **13.**d; **14.**c; **15.**a; **16.**a; **17.**a; **18.**c; **19.**c; **20.**a; **21.**b; **22.**a; **23.**c; **24.**c; **25.**b

# 13

# The Endocrine System

## Chapter Summary and Key Concepts

*After reading and studying this chapter you should know the following:*

### Sections 13.1, 13.2, 13.3

1. The endocrine system is a collection of specialized cells, tissues, and glands that produce hormones. Hormones are secreted into extracellular space and enter the blood or lymph.

2. Hormones act only on their target cells.

3. Endocrine control of body activities is slower than nervous control.

4. Hormones are classified as steroid or nonsteroid.

5. Steroid hormones are synthesized from cholesterol, are lipid soluble, and exert their effect by binding to intracellular receptors and activating or inhibiting genes.

6. Nonsteroid hormones are synthesized from amino acids, are not lipid soluble, and exert their effect by binding to a cell membrane and activating a second messenger molecule.

7. Many hormones participate in negative feedback loops that maintain homeostasis.

8. The hypothalamus is a region of the brain that interacts with the endocrine system primarily through the pituitary gland.

9. The pituitary gland consists of a posterior lobe and an anterior lobe.

10. The posterior pituitary stores ADH and oxytocin secreted by neuroendocrine cells in the hypothalamus.

11. The anterior pituitary produces ACTH, TSH, FSH, LH, PRL, and GH. Releasing and inhibiting hormones produced by the hypothalamus control the release of these hormones.

### Sections 13.4, 13.5, 13.6, 13.7, 13.8

12. The pancreas secretes insulin when blood sugar levels are high, and glucagon when blood sugar levels are low. The pancreas also secretes somatostatin, a hormone that inhibits insulin and glucagon secretion.

13. The adrenal cortex produces cortisol that helps maintain blood glucose levels, and aldosterone that helps maintain water balance. The adrenal cortex also produces small amounts of sex hormones.

14. The adrenal medulla produces epinephrine and norepinephrine. The release of these hormones is stimulated by the sympathetic nervous system, and they contribute to the fight-or-flight response.

15. The thyroid gland produces thyroxin that regulates the production of ATP, and calcitonin that decreases blood calcium levels.

16. The parathyroid glands produce parathyroid hormone that increases the blood calcium levels.

17. The testes produce testosterone and the ovaries produce estrogen and progesterone. These sex hormones regulate sexual development, activities, and secondary sex characteristics.

18. The thymus gland secretes thymosin and thymopoietin for the maturation of T cells; the pineal gland secretes melatonin that may play a role in sleep/wake cycles; and the heart, kidneys, and digestive system all secrete hormones specialized for their functions.

### Sections 13.9, 13.10

19. Histamine, prostaglandins, nitric oxide, and growth factors are chemical messengers similar to hormones but with only local effects.

20. Disorders of the endocrine system include diabetes mellitus, hyperthyroidism, hypothyroidism, Addison's disease, and Cushing's syndrome.

# Exercises

*Complete the exercises for each section after you have read and studied the section. If you cannot answer some questions, or answer them incorrectly, return to the chapter and review this information. You may find it helpful to work on only one section at a time. When you have completed all sections, take the Chapter Test as an indicator of your mastery of this topic.*

**13.1** The endocrine system produces hormones

**13.2** Hormones are classified as steroid or nonsteroid

**13.3** The hypothalamus and the pituitary gland

## Matching

_____ 1. **endocrine system**    a. an inactive molecule in a cell converted to an active molecule by the binding of a hormone to the cell membrane

_____ 2. **hormones**    b. cells in the hypothalamus that function as both nerve cells and endocrine cells

_____ 3. **endocrine glands**    c. a small region in the forebrain that plays an important role in homeostatic regulation

_____ 4. **target cells**    d. a communication system that uses chemical messengers

_____ 5. **steroid hormones**    e. a small endocrine gland called the master gland

_____ 6. **nonsteroid hormones**    f. groups of cells that secrete hormones into the extracellular space

_____ 7. **second messenger**    g. bloodborne units of information secreted by glands

_____ 8. **hypothalamus**    h. lipid-soluble hormones synthesized from cholesterol

_____ 9. **pituitary gland**    i. mostly lipid-insoluble hormones synthesized from amino acids

_____ 10. **neuroendocrine cells**    j. cells responsive to a particular hormone

## Fill-in-the-Blank

*Referenced sections are in parentheses.*

11. Hormones circulate in the _____ and act only on specific _____. (13.1)

12. There are approximately _____ different hormones. (13.1)

13. The target cells of a hormone will have the right _____ for that particular hormone. (13.1)

14. The _____ system and the _____ system interact with each other to maintain homeostasis. (13.1)

15. _____ hormones enter the cell to carry out their effects. (13.2)

16. _____ hormones may utilize a second messenger to carry out their effects. (13.2)

17. A hormone-receptor complex is involved in the action of _____ hormones. (13.2)

18. A second messenger is involved in the action of _____ hormones. (13.2)

19. In a negative feedback loop involving a hormone, the effectors are the _____ _____. (13.2)

20. In addition to producing hormones, the hypothalamus monitors the _____ _____, _____ _____, _____, and _____ _____. (13.3)

21. Neuroendocrine cells have cell bodies located in the _____ and axon endings located in the _____ _____. (13.3)

22. The connection between the hypothalamus and the posterior pituitary is _____, while the connection between the hypothalamus and the anterior pituitary is _____. (13.3)

23. Diabetes insipidus may be caused by inadequate amounts of _____. (13.3)

## Chapter 13 The Endocrine System

## Labeling

*Identify each organ in Figure 13.1.*

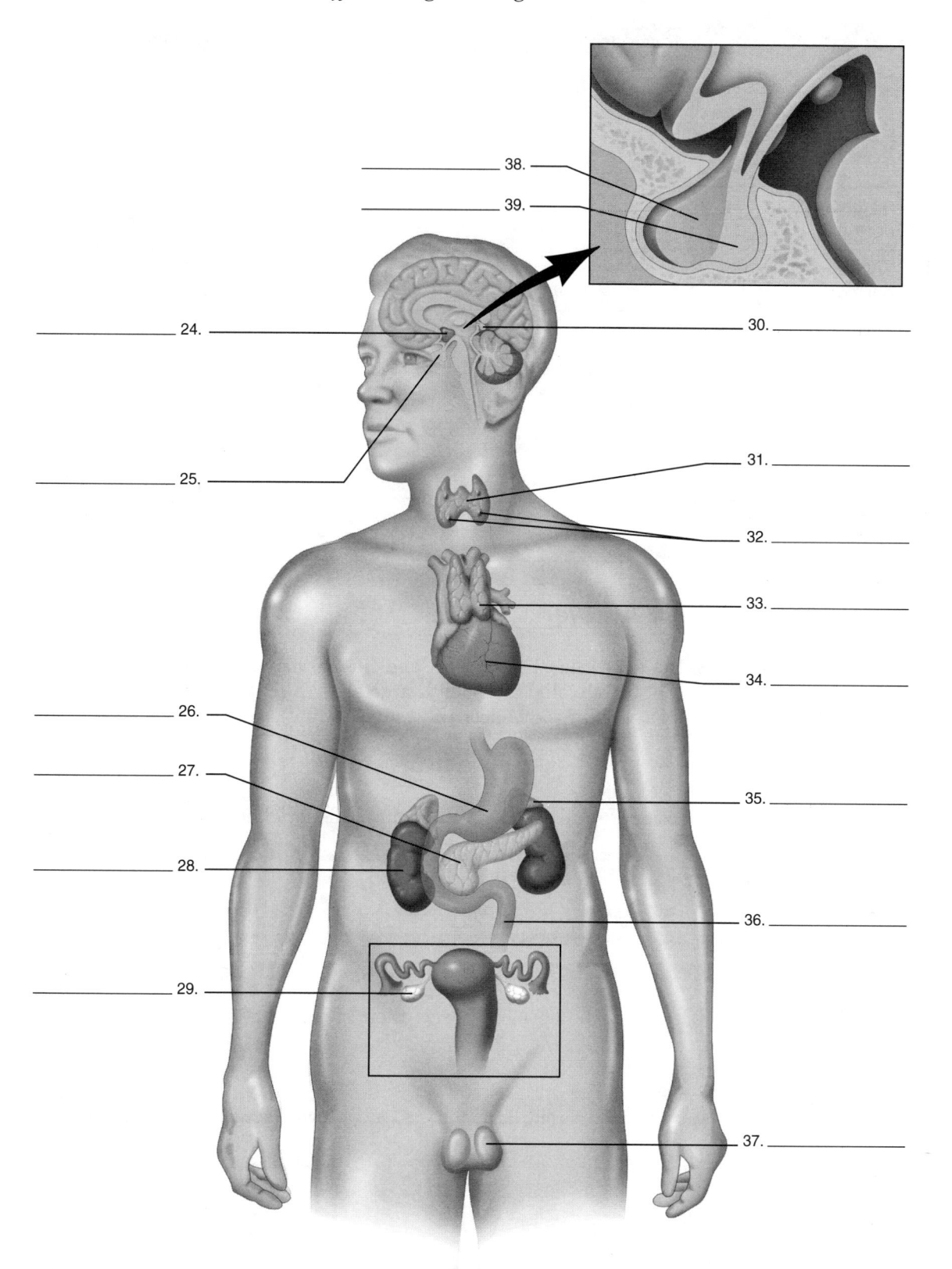

24. _____
25. _____
26. _____
27. _____
28. _____
29. _____
30. _____
31. _____
32. _____
33. _____
34. _____
35. _____
36. _____
37. _____
38. _____
39. _____

**Figure 13.1**

## Completion

*For each hormone listed below, fill in the site of production, action, and target cell.*

| Hormone | Production site | Action | Target Cell |
|---|---|---|---|
| 40. ADH | a. | b. | c. |
| 41. Oxytocin | a. | b. | c. |
| 42. Adrenocorticotropic hormone | a. | b. | c. |
| 43. Thyroid-stimulating hormone | a. | b. | c. |
| 44. Follicle-stimulating hormone | a. | b. | c. |
| 45. Leutinizing hormone | a. | b. | c. |
| 46. Prolactin | a. | b. | c. |
| 47. Growth hormone | a. | b. | c. |

## Short Answer

48. Both the endocrine system and the nervous system send messages that affect cells. List four characteristics that describe the operation of the endocrine system.

49. Compare and contrast steroid and nonsteroid hormones by describing:

    a. what chemical molecule is utilized in hormone synthesis

    b. lipid solubility

    c. how the hormones exert their effect on a target cell

50. Why are steroid hormones usually slower acting than nonsteroid hormones?

**13.4** The pancreas secretes glucagon, insulin, and somatostatin

**13.5** The adrenal gland comprises the cortex and medulla

**13.6** Thyroid and parathyroid glands

**13.7** The testes and ovaries produce sex hormones

**13.8** Other glands and organs also secrete hormones

## Matching

*Match each hormone with the correct function.*

_____ 1. **glucagon**

_____ 2. **insulin**

_____ 3. **somatostatin**

_____ 4. **cortisol**

_____ 5. **aldosterone**

_____ 6. **epinephrine**

_____ 7. **thyroxine**

_____ 8. **calcitonin**

_____ 9. **parathyroid hormone**

_____ 10. **testosterone**

_____ 11. **estrogen**

_____ 12. **thymosin**

_____ 13. **melatonin**

_____ 14. **atrial natriuretic hormone**

_____ 15. **cholecystokinin**

_____ 16. **erythropoietin**

_____ 17. **renin**

a. contributes to the action of the sympathetic nervous system

b. stimulates the production of red blood cells in the bone marrow

c. helps lymphocytes of the immune system mature into T cells

d. lowers blood calcium levels by increasing the deposition of calcium in bone

e. helps to regulate blood pressure

f. raises blood sugar levels

g. regulates the development of sperm, the male reproductive organs, and male secondary sex characteristics

h. lowers blood sugar levels

i. aids the functioning of the digestive system

j. a glucocorticoid that helps maintain blood sugar levels by promoting the breakdown of fats and proteins

k. stimulates aldosterone secretion and constricts blood vessels

l. a mineralocorticoid that regulates the amount of sodium and potassium in the body

m. inhibits the secretion of glucagon and insulin

n. regulates the production of ATP in body cells

o. initiates development of female secondary sex characteristics and helps regulate the menstrual cycle

p. increases blood calcium levels

q. may help synchronize body rhythms to the light/dark cycle

## Crossword Puzzle

Write in the hormone that will be secreted in response to each physiological situation represented in each clue.

### Across

1. _____ hormone is secreted when blood calcium levels are too low
3. Blood calcium levels are too high
5. A sudden crisis activates the sympathetic nervous system
7. Injury and emotional stress
9. _____ natriuretic hormone secreted when blood pressure rises

### Down

2. Potassium levels in the body are too high
4. Blood sugar levels are too low
6. Blood glucose levels increase
8. A situation that creates an increased need for energy
10. Iodine deficiency leads to absence of thyroxin

**13.9** Other chemical messengers

**13.10** Disorders of the endocrine system

# Completion

*Complete the table below, writing in the action associated with each chemical messenger.*

| Chemical Messenger | Primary Action |
|---|---|
| histamine | 1. |
| prostaglandins | 2. |
| nitric oxide | 3. |
| growth factor | 4. |

*For each of the endocrine disorders listed below, write in the cause and symptoms.*

5. Hypothyroidism

   a. cause:

   b. symptoms:

6. Hyperthyroidism

   a. cause:

   b. symptoms:

7. Addison's disease

   a. cause:

   b. symptoms:

8. Cushing's syndrome
   a. cause:

   b. symptoms:

9. Type I diabetes
   a. cause:

   b. symptoms:

10. Type II diabetes
    a. cause:

    b. symptoms:

---

# Chapter Test

## Multiple Choice

1. Endocrine glands:
   a. secrete hormones by way of a duct into the area near a target organ.
   b. secrete mucus, sweat, and digestive fluids.
   c. contain specialized nerve cells that store pre-hormone molecules for synthesis.
   d. are ductless organs.

2. The target cells of an endocrine gland:
   a. have specific receptors for the hormone.
   b. respond to all endocrine hormones released by all endocrine glands.
   c. will always be in close proximity to the gland.
   d. are connected to the gland by nerve cells.

3. Endocrine control of body activities is usually _____ than nervous control.
   a. faster
   b. slower
   c. more specific
   d. less specific

4. The nervous system and the endocrine system interact through the:
   a. thyroid gland and the pituitary gland.
   b. pituitary gland and the hypothalamus.
   c. hypothalamus and the thyroid gland.
   d. anterior pituitary and posterior pituitary.

5. Steroid hormones exert their effect using:
   a. a second messenger.
   b. cyclic AMP.
   c. a hormone-receptor complex.
   d. a carbohydrate molecule.

6. Nonsteroid hormones:
   a. cross the cell membrane easily.
   b. bind to receptors on the cell surface.
   c. cross the membrane or bind to receptors, depending on the target cell.
   d. pass through protein channels in the membrane.

7. Nonsteroid hormones are usually faster acting than steroid hormones because:
   a. they are transported more quickly.
   b. they activate proteins that already exist within the cell.
   c. they utilize enzymes.
   d. they are not attacked by immune cells.

8. The hypothalamus is able to produce hormones because:
   a. it is in the brain.
   b. it contains neurosecretory cells.
   c. it is influenced by the pituitary gland.
   d. it is connected to the thalamus, which functions as a gland.

9. All of the following are accurate descriptions of the hypothalamus *except:*
   a. neuroendocrine cells in the hypothalamus store hormones in the posterior pituitary.
   b. ADH stimulates the hypothalamus to produce oxytocin.
   c. the hypothalamus contains the cell bodies of neuroendocrine cells.
   d. the hypothalamus functions in both the nervous system and the endocrine system.

10. Which of the following hormones is *not* produced by the anterior pituitary?
    a. adrenocorticotropic hormone
    b. luteinizing hormone
    c. prolactin
    d. antidiuretic hormone

11. Hormones produced in the anterior pituitary will be released in response to _____ produced by the _____.
    a. factors, thyroid
    b. releasing hormones, anterior pituitary
    c. releasing hormones, posterior pituitary
    d. releasing hormones, hypothalamus

12. FSH stimulates:
    a. the release of growth hormone by the pituitary gland.
    b. egg development in females and sperm development in males.
    c. ovulation in females and testosterone production in males.
    d. uterine contraction and milk production in pregnant or lactating females.

13. Diabetes insipidus:
    a. results in decreased production of insulin.
    b. is caused by a virus.
    c. is characterized by excessive water loss through the kidneys.
    d. is caused by lack of ADH or poor response to ADH.
    e. a and c
    f. c and d

14. The _____ is both an endocrine and an exocrine gland.
    a. thyroid gland
    b. pituitary gland
    c. hypothalamus
    d. pancreas

15. Which of the following hormones will be most active after a 24-hour fast?
    a. insulin
    b. aldosterone
    c. glucagon
    d. somatostatin

16. Cortisol production occurs in the _____ and is stimulated by _____.
    a. adrenal medulla, FSH
    b. adrenal cortex, ACTH
    c. thyroid gland, TSH
    d. thymus gland, growth hormone

17. The adrenal medulla is similar to the hypothalamus because:
    a. they are both neuroendocrine organs.
    b. they are both divided into two regions.
    c. they both produce inhibiting and releasing hormones.
    d. they both respond to pituitary hormones.

18. The blood calcium regulatory function of which gland is most important in adulthood?
    a. thyroid gland
    b. parathyroid gland
    c. pancreas
    d. thymus gland

19. Goiter is caused by:
    a. increased levels of thyroxin.
    b. iodine deficiency.
    c. calcium deficiency.
    d. malfunction of the hypothalamus.

20. Which of the following is *not* a characteristic of parathyroid hormone?
    a. It increases blood calcium levels.
    b. It inhibits calcium retention by the kidneys.
    c. It increases calcium reabsorption by the digestive tract.
    d. It removes calcium from the bone.

21. Testosterone:
    a. is produced by the anterior pituitary.
    b. is produced at consistent levels throughout life.
    c. secretion is stimulated by LH.
    d. inhibits the production of aldosterone.

22. The kidneys secrete:
    a. secretin and cholecystokinin.
    b. renin and erythropoietin.
    c. histamine and prostaglandins.
    d. aldosterone and ADH.

23. Nitric oxide is not considered a hormone because:
    a. it is not a steroid.
    b. it is short lived.
    c. it is produced by epithelial tissue.
    d. it is not secreted into the bloodstream.

24. Which of the following sets of hormones oppose each other?
    a. secretin and renin
    b. FSH and LH
    c. glucagon and insulin
    d. progesterone and estrogen

25. The symptoms of Cushing's disease are caused by:
    a. lowered levels of cortisol and aldosterone.
    b. hypersecretion by the thyroid gland.
    c. excessive levels of cortisol.
    d. excessive levels of testosterone.

# Key Concept Review Questions

*Each of the Key Concepts listed at the beginning of this chapter has been rewritten as a question below. After successfully completing the study guide exercises and the Chapter Test, you should be able to answer each of these questions. Refer to the Key Concepts list at the beginning of this chapter to check your answers.*

1. What structures are included in the endocrine system? What is the initial destination of secreted hormones?
2. On what cells do hormones act?
3. How does endocrine control of body activities compare to nervous control in terms of speed?
4. What are the two general classifications of hormones?
5. What are the characteristics of steroid hormones?
6. What are the characteristics of nonsteroid hormones?
7. What type of feedback loops do hormones participate in?
8. What gland allows interaction between the hypothalamus and the endocrine system?
9. What are the two regions of the pituitary gland?
10. What hormones are stored in the posterior pituitary? Where were these hormones produced?
11. What are the hormones produced by the anterior pituitary? What controls the release of these hormones?
12. What are the three hormones produced by the pancreas? What is the function of each hormone?

13. What hormones are produced by the adrenal cortex? What are the functions of these hormones?

14. What hormones are produced by the adrenal medulla? What stimulates release of these hormones?

15. What hormones are produced by the thyroid gland? What are the functions of these hormones?

16. What hormone is produced by the parathyroid gland? What is the function of this hormone?

17. What hormones are produced by the testes and the ovaries? What are the functions of these hormones?

18. What are the names and functions of the hormones produced by the thymus gland and pineal gland?

19. What are the names of four chemical messengers?

20. What are at least four disorders of the endocrine system discussed in this chapter?

## Answer Key

### Sections 13.1, 13.2, 13.3

**1.** d; **2.** g; **3.** f; **4.** j; **5.** h; **6.** i; **7.** a; **8.** c; **9.** e; **10.** b; **11.** bloodstream, cells; **12.** 50; **13.** receptors; **14.** endocrine, nervous; **15.** Steroid; **16.** Nonsteroid; **17.** steroid; **18.** nonsteroid; **19.** target cells; **20.** pituitary gland, internal environment, temperature, carbohydrate metabolism; **21.** hypothalamus, posterior pituitary; **22.** neural, endocrine; **23.** ADH; **24.** hypothalamus; **25.** pituitary gland; **26.** stomach; **27.** pancreas; **28.** kidneys; **29.** ovaries; **30.** pineal gland; **31.** thyroid gland; **32.** parathyroid gland; **33.** thymus gland; **34.** heart; **35.** adrenal glands; **36.** intestines; **37.** testes; **38.** anterior lobe of the pituitary; **39.** posterior lobe of the pituitary; **40.** a. hypothalamus, b. regulates water balance, c. kidney; **41.** a. hypothalamus, b. stimulates uterine contraction and milk ejection, c. uterus and mammary glands; **42.** a. anterior pituitary, b. stimulates release of glucocorticoids, c. adrenal cortex; **43.** a. anterior pituitary, b. stimulates synthesis and release of thyroid hormone, c. thyroid gland; **44.** a. anterior pituitary, b. egg development in females and sperm development in males, c. reproductive organs; **45.** a. anterior pituitary, b. ovulation in females and testosterone production in males, c. reproductive organs; **46.** a. anterior pituitary, b. stimulates milk production, c. mammary glands; **47.** a. anterior pituitary, b. promotes cell growth, c. widespread effect on many cells; **48.** endocrine hormones reach almost all cells, a hormone acts only on its target cell, endocrine system messages are slower than nervous system messages, and the endocrine system interacts with the nervous system; **49.** a. cholesterol is used in the synthesis of steroid hormones, while nonsteroid hormones utilize amino acids, b. steroid hormones are lipid soluble, while nonsteroid hormones are not, c. steroid hormones exert an effect after they enter a target cell, while nonsteroid hormones bind to the target cell surface and trigger a series of events that affect the target cell; **50.** Steroid hormones act by affecting DNA activity and increasing or decreasing production of a critical protein; because non-steroid hormones affect proteins that have already been synthesized, their effect is faster.

## Sections 13.4, 13.5, 13.6, 13.7, 13.8

**1.**f; **2.**h; **3.**m; **4.**j; **5.**l; **6.**a; **7.**n; **8.**d; **9.**p; **10.**g; **11.**o; **12.**c; **13.**q; **14.**e; **15.**i; **16.**b; **17.**k; **Crossword Puzzle: 1.** Parathyroid; **2.** aldosterone; **3.** calcitonin; **4.** glucagons; **5.** epinephrine; **6.** insulin; **7.** cortisol; **8.** thyroxin; **9.** Atrial; **10.** TSH

## Sections 13.9, 13.10

**1.** increases mucus secretion, dilates blood vessels, and increases the leakiness of capillaries; **2.** locally controls blood flow, contributes to inflammatory response, aids blood clotting; **3.** regulates smooth muscle contraction in the digestive tract, controls penile erection; **4.** influences cell division and the direction of cell growth; **5.** a. varied, b. slow growth in children, adults may have swelling, lethargy, and weight gain; **6.** a. may be an autoimmune disorder, b. includes protruding eyes; **7.** a. decreased secretion of cortisol and aldosterone by the adrenal cortex, b. fatigue, weakness, abdominal pain, weight loss, bronzed skin color; **8.** a. excessive cortisol production, may be due to a tumor, b. muscle weakness and fatigue, swelling, high blood pressure; **9.** a. pancreas cannot produce enough insulin, may be caused by a virus, b. dehydration, thirst, fatigue, frequent urination, blurred vision, slow wound healing; **10.** a. cells are unable to respond to insulin, caused by genetics and lifestyle; b. symptoms similar to those for type I diabetes

## Chapter Test

**1.**d; **2.**a; **3.**b; **4.**b; **5.**c; **6.**b; **7.**b; **8.**b; **9.**b; **10.**d; **11.**d; **12.**b; **13.**f; **14.**d; **15.**c; **16.**b; **17.**a; **18.**b; **19.**b; **20.**b; **21.**c; **22.**b; **23.**d; **24.**c; **25.**c

# 14

# The Digestive System and Nutrition

## Chapter Summary and Key Concepts

*After reading and studying this chapter you should know the following:*

### Sections 14.1, 14.2, 14.3, 14.4, 14.5

1. The digestive system provides nutrients to the body.
2. The walls of the gastrointestinal tract consist of four layers: the innermost mucosa, the submucosa, the muscularis, and the serosa.
3. Sphincters are thick rings of ciliary muscle that can close off the passageways between organs.
4. The five basic processes of the digestive system are mechanical processing and movement, secretion, digestion, absorption, and elimination.
5. Two types of motility important to the digestive system are peristalsis that moves food forward, and segmentation that mixes food.
6. In the mouth the teeth grind the food, and saliva begins the process of digestion.
7. The pharynx receives food from the mouth and initiates swallowing.
8. The esophagus is a muscular tube that connects the pharynx and the stomach.
9. The stomach functions to store food, digest food, and regulate the delivery of food to the small intestine.
10. The small intestine functions to digest carbohydrates, lipids, and proteins, and to absorb most of the nutrients.

### Sections 14.6, 14.7, 14.8, 14.9

11. The accessory organs of the digestive system include the salivary glands, the pancreas, the liver, and the gallbladder.

12. The pancreas secretes digestive enzymes that complete digestion in the small intestine, and sodium bicarbonate that neutralizes stomach acid.

13. The liver produces bile, stores fat-soluble vitamins, stores glucose as glycogen, manufactures plasma proteins, synthesizes and stores some lipids, inactivates chemicals, converts ammonia to urea, and destroys worn-out red blood cells.

14. The gallbladder stores and releases bile.

15. The large intestine reabsorbs water and concentrates feces.

16. Proteins and carbohydrates are absorbed from the small intestine by active transport, lipids move by diffusion, and water moves by osmosis.

17. Digestion is regulated by the endocrine system and the nervous system.

### Sections 14.10, 14.11, 14.12, 14.13

18. Carbohydrates, lipids, and proteins are essential nutrients in the diet.

19. Body weight depends on the balance between energy consumed and energy used.

20. Disorders of the digestive system include lactose intolerance, diverticulosis, colon polyps, hepatitis, gallstones, malnutrition, anorexia nervosa, and bulimia.

# Exercises

Complete the exercises for each section after you have read and studied the section. If you cannot answer some questions, or answer them incorrectly, return to the chapter and review this information. You may find it helpful to work on only one section at a time. When you have completed all sections, take the Chapter Test as an indicator of your mastery of this topic.

**14.1   The digestive system brings nutrients into the body**

**14.2   The mouth processes food for swallowing**

**14.3   The pharynx and esophagus deliver food to the stomach**

**14.4   The stomach stores, digests protein, and regulates delivery**

**14.5   The small intestine digests food and absorbs nutrients**

## Matching

_e_ 1. **digestive system**
_j_ 2. **gastrointestinal tract**
_q_ 3. **accessory organs**
_h_ 4. **peristalsis**
_o_ 5. **segmentation**
_r_ 6. **mouth**
_a_ 7. **salivary glands**
_m_ 8. **pharynx**
_n_ 9. **esophagus**
_p_ 10. **stomach**
_b_ 11. **gastric glands**
_g_ 12. **pepsin**
_i_ 13. **gastric juice**
_d_ 14. **chyme**
_k_ 15. **small intestine**
_c_ 16. **duodenum**
_l_ 17. **villi**
_f_ 18. **lacteal**

a. three glands near the oral cavity that produce saliva
b. secretory cells in the mucosa of the stomach
c. the region of the small intestine where most digestion occurs
d. a watery mixture of partially digested food and gastric juice that enters the small intestine
e. a body system consisting of all the organs that function in getting nutrients into the body
f. capillaries and a lymph vessel at the center of a villus, which function to transport nutrients to larger vessels
g. a protein-digesting enzyme in the stomach
h. muscular waves of contraction that propel food forward
i. a mixture secreted by the stomach that contains HCl, pepsinogen, and fluid
j. a hollow tube formed by the hollow organs of the digestive system
k. an organ that receives chyme from the stomach, secretions from the pancreas, and bile from the gallbladder
l. microscopic projections that cover large folds in the mucosa of the small intestine
m. the throat
n. a muscular tube that connects the pharynx to the stomach
o. random contraction and relaxation of smooth muscle that mixes food
p. a muscular sac that stores and digests food and regulates food delivery to the small intestine
q. organs that aid the digestive system and are not part of the gastrointestinal tract
r. location where digestion begins

## Labeling

*Label each indicated structure in Figure 14.1 with a name and a function.*

**Structures**
a. stomach
b. pancreas
c. mouth
d. liver
e. pharynx

**Functions**
m. secretes saliva
n. food is chewed and moistened here
o. swallowing is initiated here
p. food mixes with gastric juice and protein digestion begins
q. produces bile

**Structures**
f. rectum
g. salivary glands
h. esophagus
i. small intestine
j. large intestine
k. anus
l. gallbladder

**Functions**
r. reabsorbs water and concentrates undigested food into feces
s. terminal opening for expelling feces
t. stores and concentrates bile
u. organ where digestion of all food is completed
v. secretes enzymes to digest all food groups
w. muscular tube that moves food from pharynx to stomach
x. passageway for fecal material

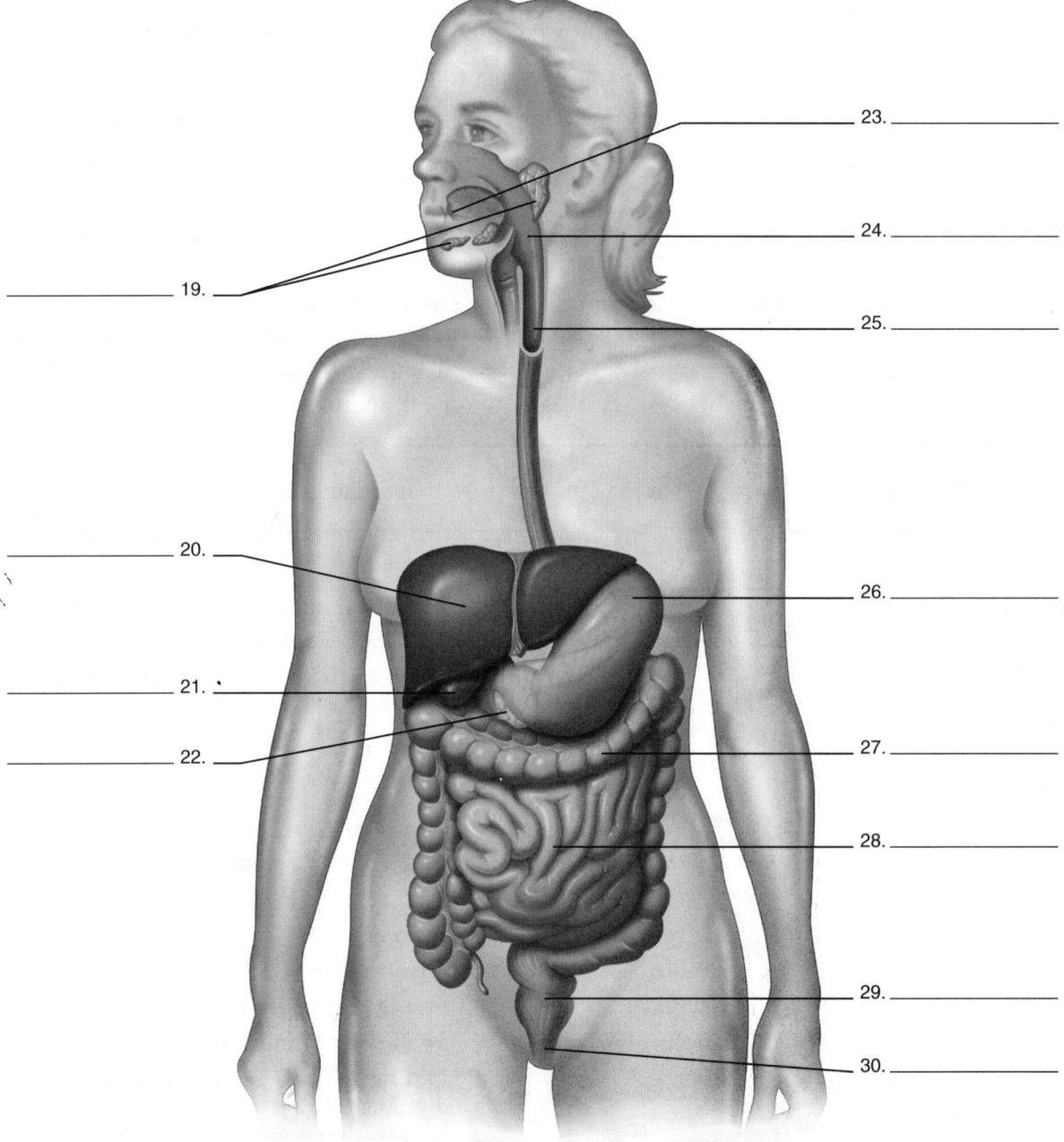

19. _____
20. _____
21. _____
22. _____
23. _____
24. _____
25. _____
26. _____
27. _____
28. _____
29. _____
30. _____

**Figure 14.1**

## Fill-in-the-Blank

*Referenced sections are in parentheses.*

31. The space within the hollow tube of the GI tract is the _____. (14.1)

32. The muscularis consists of _____ muscle in a _____ and a _____ arrangement. (14.1)

33. The _____ is the entrance to the GI tract. (14.2)

34. The type of tooth specialized for cutting food is the _____. (14.2)

35. The visible region of a tooth is the _____. (14.2)

36. _____ is an enzyme found in saliva that begins the digestion of carbohydrates. (14.2)

37. Once initiated, the swallowing reflex is _____ and cannot be stopped. (14.3)

38. The _____ _____ is a circular muscle that lies between the stomach and the small intestine. (14.4)

39. Gastric glands release _____ _____ which consists of HCL, pepsinogen, and fluid. (14.4)

40. Stretching in the stomach signals the muscles to increase _____. (14.4)

41. The first region of the small intestine is the _____. (14.5)

42. The _____ and the _____ are segments of the small intestine where most absorption takes place. (14.5)

43. The surface of each epithelial cell of a villus is covered with smaller projections called _____. (14.4)

## Labeling

*Label each indicated structure in Figure 14.2 and then answer the questions below.*

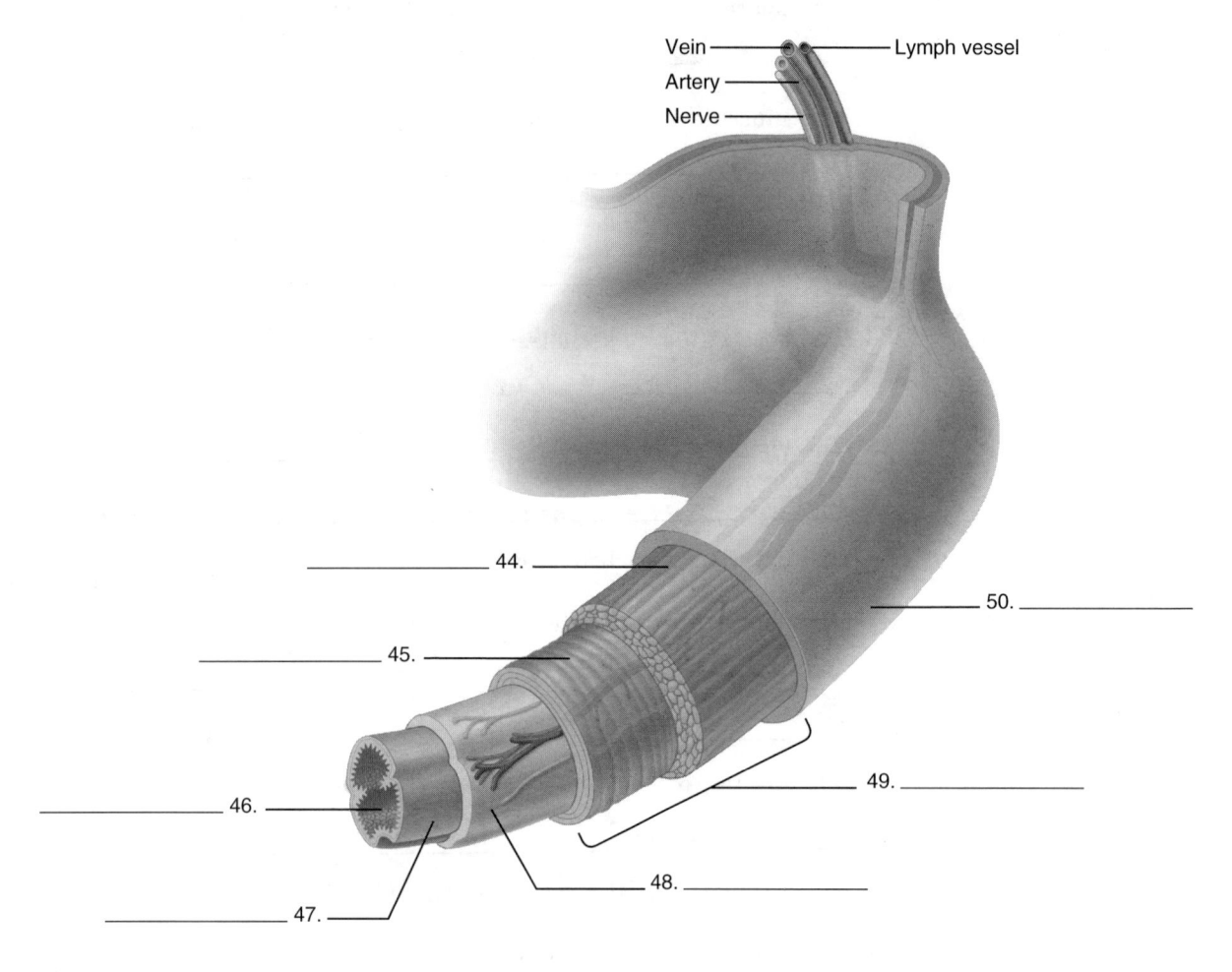

Vein
Lymph vessel
Artery
Nerve

44. _____
45. _____
46. _____
47. _____
48. _____
49. _____
50. _____

**Figure 14.2**

51. Which layer is responsible for motility of the digestive tract?
52. Which layer carries blood vessels and nerves?
53. Which layer protects and anchors the digestive tract?
54. Which layer contains smooth muscle?

14.6 Accessory organs aid digestion and absorption

14.7 The large intestine absorbs nutrients and eliminates waste

14.8 How nutrients are absorbed

14.9 Endocrine and nervous systems regulate digestion

## Matching

*Write the letter(s) of each function next to the correct organs.*

_____ 1. **liver**

_____ 2. **pancreas**

_____ 3. **gallbladder**

_____ 4. **colon**

_____ 5. **cecum**

_____ 6. **large intestine**

_____ 7. **stomach**

a. is divided into ascending, transverse, and descending sections

b. stores fat-soluble vitamins

c. secretes sodium bicarbonate to maintain the more neutral pH required for its function

d. stores excess glucose as glycogen

e. detoxifies substances

f. a pouch that receives chime from the small intestine

g. absorbs water and produces solid waste

h. produces bile

i. receives nutrients delivered by the hepatic portal system

j. secretes proteases to digest protein and amylase to digest carbohydrates

k. concentrates and stores bile

l. breaks down old red blood cells

m. stretching increases peristalsis

n. delivers digestive enzymes to the duodenum by way of two ducts

o. cholecystokinin (CCK) is produced and stimulates the release of bile

## Paragraph Completion

*Use the terms below to complete the paragraph, about nutrient absorption, then repeat this exercise with the terms covered.*

| | | | |
|---|---|---|---|
| capillaries | disaccharide | mouth | facilitated diffusion |
| monosaccharides | small intestine | micelles | chylomicrons |
| active transport | small intestine | stomach | exocytosis |
| monoglycerides | nonpolar | polar | amino acids |
| small intestine | fatty acids | lymph | triglycerides |

Protein digestion begins in the (8) _____ and is completed in the (9) _____ _____. This final breakdown of protein produces (10) _____ _____ that are moved by the process of (11) _____ _____ into mucosal cells of the intestine. They will eventually leave the mucosal cells by the process of (12) _____ _____ and move into the (13) _____.

Carbohydrate digestion begins in the (14) _____ where polysaccharides are broken down into (15) _____. Carbohydrate digestion is completed in the (16) _____ _____ where (17) _____ are produced and transported into the capillaries.

Lipids are digested in the (18) _____ _____ where they are broken down into (19) _____ _____ and (20) _____. These components are nonpolar and dissolve in (21) _____ that have a (22) _____ outer surface and a (23) _____ inner core. They are then transported inside mucosal cells, where they recombine to form (24) _____. Clusters of triglycerides are coated with proteins and form (25) _____ which leave the mucosal cells by the process of (26) _____. Instead of entering capillaries, they enter and travel through (27) _____ vessels.

14.10    Nutrition: You are what you eat
14.11    Weight control: Energy consumed versus energy spent
14.12    Disorders of the digestive system
14.13    Eating disorders: Anorexia nervosa and bulimia

## Short Answer

1. Why is it healthier to eat complex carbohydrates than simple sugars?

2. What foods provide saturated fats to the diet? Why are saturated fats considered to be unhealthy?

3. Why are eight of the amino acids used in protein synthesis called "essential amino acids"?

4. What is a complete protein? What foods provide complete protein?

5. What percent of daily calories should come from each of the three main nutrient groups?

6. Why is it unnecessary to take supplements that contain fat-soluble vitamins?

7. What is fiber and why is it not considered a nutrient source?

8. Under of the following conditions, will the BMR increase, decrease, or remain the same?
    a. gender
    b. stress
    c. eating
    d. not eating
    e. obesity

## Completion

*Indicate the correct digestive system or eating disorder, cause, and symptoms by completing the table below.*

| Disorder | Cause | Symptoms |
|---|---|---|
| Lactose intolerance | 9. | 10. |
| 11. | small protrusion of the intestinal wall, may be caused by a low-fiber diet | sometimes has no symptoms |
| 12. | liver inflammation caused by contaminated food or water | brief illness |
| Hepatitis B | 13. | 14. |
| Hepatitis C | 15. | 16. |
| 17. | High levels of cholesterol in bile | 18. |
| 19. | excessive dieting, cause unknown | 20. |
| Bulemia | 21. | 22. |

# Chapter Test

## Multiple Choice

1. The gastrointestinal tract is composed of:
   a. all the organs of the digestive system.
   b. the accessory organs of the digestive system.
   c. the hollow organs of the digestive system.
   d. the blood vessels and nerves that supply the digestive system.

2. Which layer of the gastrointestinal tract wall is connective tissue containing blood vessels, lymph vessels, and nerves?
   a. mucosa
   b. submucosa
   c. serosa
   d. muscularis

3. Food is propelled forward through the GI tract by:
   a. mechanical processing.
   b. peristalsis.
   c. segmentation.
   d. motility.

4. The innermost region of a tooth containing blood vessels and nerves is the:
   a. enamel.
   b. crown.
   c. pulp cavity.
   d. dentin.

5. The tongue is primarily:
   a. skeletal muscle.
   b. smooth muscle.
   c. connective tissue.
   d. epithelial tissue.

6. Which of the following is *not* a component of saliva?
   a. mucin
   b. bicarbonate
   c. lysozyme
   d. proteinase

7. Which of the following represents the path taken by food after leaving the mouth?
   a. pharynx, larynx, esophagus, small intestine, large intestine
   b. pharynx, esophagus, stomach, small intestine, large intestine
   c. larynx, esophagus, stomach, large intestine, small intestine
   d. larynx, esophagus, stomach, small intestine, large intestine

8. Malfunction of the lower esophageal sphincter may result in:
   a. bulimia.
   b. acid reflux.
   c. stomach ulcer.
   d. diarrhea.

9. Gastric glands secrete:
   a. HCl and pepsinogen.
   b. gastrin and secretin.
   c. gastrin and lipases.
   d. intrinsic factor and amylase.

10. Enzyme activity in the stomach contributes to the digestion of:
    a. carbohydrates.
    b. lipids.
    c. proteins.
    d. all of the above

11. Most of the digestion that takes place in the small intestine occurs in the:
    a. duodenum.
    b. ileum.
    c. jejunum.
    d. lacteals.

12. The surface area of the small intestine is greatly increased by the presence of:
    a. serosa and submucosa.
    b. muscularis and epithelial cells.
    c. duodenum, jejunum, and ileum.
    d. villi and microvilli.

13. The pancreas secretes enzymes that aid the digestion of:
    a. carbohydrates.
    b. lipids.
    c. proteins.
    d. all of the above

14. Bile is produced by the _____ and stored in the _____.
    a. gallbladder, small intestine
    b. gallbladder, liver
    c. liver, small intestine
    d. liver, gallbladder

15. The hepatic portal system delivers nutrient-rich blood from the _____ to the _____.
    a. GI tract, liver
    b. liver, small intestine
    c. stomach, liver
    d. liver, gall bladder

16. The large intestine absorbs primarily:
    a. nutrients.
    b. vitamins and minerals.
    c. lipids.
    d. water.

17. Which of the following represents the correct sequence in which structures involved in lipid absorption are formed?
    a. micelles, triglycerides, chylomicrons
    b. triglycerides, chylomicrons, micelles
    c. micelles, chylomicrons, triglycerides
    d. chylomicrons, micelles, triglycerides

18. Proteins and carbohydrates are absorbed from the small intestine by:
    a. diffusion.
    b. facilitated diffusion.
    c. active transport.
    d. osmosis.

19. Most digestive regulatory processes involve the:
    a. mouth and stomach.
    b. stomach and large intestine.
    c. stomach and small intestine.
    d. small intestine and large intestine.

20. Complex carbohydrates are better nutritionally than simple carbohydrates because:
    a. simple carbohydrates contain more fats.
    b. simple carbohydrates contain more calories.
    c. complex carbohydrates provide a more lasting source of energy.
    d. complex carbohydrates are easier to absorb.

21. Essential amino acids are those that:
    a. are required for protein synthesis.
    b. the body is capable of synthesizing.
    c. the body is not capable of synthesizing.
    d. are quickly broken down in the body.

22. Water soluble vitamins _____ stored in the liver and _____ get consumed daily.
    a. are, do not
    b. are, do
    c. are not, do not
    d. are not, do

23. Your basal metabolic rate is:
    a. the number of calories you consume each day.
    b. the number of calories you use each day.
    c. the number of calories your body needs to perform essential activities.
    d. the number of calories your body needs to perform nonessential activities.

24. To lose 1 pound of fat, you must spend:
    a. 1,000 calories.
    b. 1,500 calories.
    c. 2,500 calories.
    d. 3,500 calories.

25. A form of hepatitis transmitted by contaminated food or water is:
    a. hepatitis A.
    b. hepatitis B.
    c. hepatitis C.
    d. hepatitis E.

# Key Concept Review Questions

*Each of the Key Concepts listed at the beginning of this chapter has been rewritten as a question below. After successfully completing the study guide exercises and the Chapter Test, you should be able to answer each of these questions. Refer to the Key Concepts list at the beginning of this chapter to check your answers.*

1. What is the function of the digestive system?
2. What are the four layers in the wall of the gastrointestinal tract?
3. What are sphincters composed of, and what is their function?
4. What are the five basic processes of the digestive system?
5. What are two types of motility that function in the digestive system?

6. What occurs in the mouth?
7. What happens in the pharynx?
8. What two organs are connected by the esophagus?
9. What are the functions of the stomach?
10. What are the functions of the small intestine?
11. What are the four accessory organs of the digestive system?
12. What are the functions of the pancreas?
13. What are the functions of the liver?
14. What are the functions of the gallbladder?
15. What are the functions of the large intestine?
16. What transport processes are used in the absorption of carbohydrates, proteins, lipids, and water?
17. What two body systems are involved in the regulation of the digestive system?
18. What are the names of three classes of compounds that are essential nutrients in the diet?
19. How does energy consumption and energy usage affect body weight?
20. What are eight disorders of the digestive system discussed in this chapter?

## Answer Key

### Sections 14.1, 14.2, 14.3, 14.4, 14.5

**1.** e; **2.** j; **3.** q; **4.** h; **5.** o; **6.** r; **7.** a; **8.** m; **9.** n; **10.** p; **11.** b; **12.** g; **13.** i; **14.** d; **15.** k; **16.** c; **17.** l; **18.** f; **19.** g,m; **20.** d,q; **21.** l,t; **22.** b,v; **23.** c; **24.** e,o; **25.** h,w; **26.** a,p; **27.** i,u; **28.** j,r; **29.** f,x; **30.** k,s; **31.** lumen; **32.** smooth, circular, lengthwise; **33.** mouth; **34.** incisor; **35.** crown; **36.** Amylase; **37.** involuntary; **38.** pyloric sphincter; **39.** gastric juice; **40.** peristalsis; **41.** duodenum; **42.** jejunum, ileum; **43.** microvilli; **44.** longitudinal layer of smooth muscle; **45.** circular layer of smooth muscle; **46.** lumen; **47.** mucosa; **48.** submucosa; **49.** muscularis; **50.** serosa; **51.** muscularis; **52.** submucosa; **53.** serosa; **54.** muscularis

### Sections 14.6, 14.7, 14.8, 14.9

**1.** m,o; **2.** b,d,e,h,i,l; **3.** c,j,n; **4.** k; **5.** a; **6.** f; **7.** g; **8.** stomach; **9.** small intestine; **10.** amino acid; **11.** active transport; **12.** facilitated diffusion; **13.** capillaries; **14.** mouth; **15.** disaccharides; **16.** small intestine; **17.** monosaccharides; **18.** small intestine; **19.** fatty acids; **20.** monoglycerides; **21.** micelles; **22.** polar; **23.** nonpolar; **24.** triglycerides; **25.** chylomicrons; **26.** exocytosis; **27.** lymph

## Sections 14.10, 14.11, 14.12, 14.13

**1.** Complex carbohydrates consist of many sugar units linked together, while simple sugars are much smaller. Complex carbohydrates release these sugar units slowly, and also provide additional healthy components.; **2.** Saturated fats are found primarily in meat and dairy products; they are unhealthy because they increase LDLs in the blood and contribute to heart disease.; **3.** Essential amino acids are those that the body cannot manufacture and must be obtained through the diet.; **4.** A complete protein contains all 20 amino acids; most animal proteins are complete protein.; **5.** carbohydrates: 45–65%, lipids: 20–35%, proteins: 15%; **6.** Fat soluble vitamins are stored in the body's fat tissues and released over time, so they don't need to be consumed daily.; **7.** Fiber is a material found in vegetables, fruits, and grains; it does not provide energy because we cannot break it down.; **8.** a. higher in males; b. increase; c. increase; d. decrease; e. decrease; **9.** loss of the enzyme that digest the mile sugar lactose; **10.** diarrhea, gas, bloating, cramps; **11.** Diverticulosis; **12.** Hepatitis A; **13.** caused by a virus passed through contaminated needles, blood, or through sexual contact; **14.** jaundice, nausea, fatigue, abdominal pain, arthritis; **15.** caused by a virus passed through contaminated needles or blood, or through sexual contact; **16.** may have no early symptoms, but may result in liver damage later; **17.** Gallstones; **18.** may have no symptoms, may cause severe pain; **19.** Anorexia nervosa; **20.** extra weight loss, fatigue, hair loss, eventual osteoporosis; **21.** unknown, symptoms caused by cycle of eating and intentional vomiting; **22.** ulcers, heartburn, rectal bleeding, gum and tooth damage

## Chapter Test

**1.**c; **2.**b; **3.**c; **4.**d; **5.**a; **6.**d; **7.**b; **8.**b; **9.**a; **10.**c; **11.**a; **12.**d; **13.**d; **14.**d; **15.**a; **16.**d; **17.**a; **18.**c; **19.**c; **20.**c; **21.**c; **22.**d; **23.**c; **24.**d; **25.**a

# 15

# The Urinary System

## Chapter Summary and Key Concepts

*After reading and studying this chapter you should know the following:*

**Sections 15.1, 15.2, 15.3**

1. Excretion refers to processes that remove wastes and excess materials from the body.

2. Excretion is accomplished by the digestive system, the lungs, the skin, the liver, and the urinary system.

3. The urinary system consists of the kidneys, ureters, bladder, and urethra.

4. The urinary system maintains a constant internal environment by regulating water balance and body levels of nitrogenous wastes, ions, and other substances.

5. The main organs of the urinary system are the kidneys.

6. The kidneys form urine and consist of the outer cortex, the inner medulla, and the renal pelvis.

7. The ureters transport urine to the bladder, the bladder stores urine, and the urethra carries urine from the body.

8. The nephron is the functional unit of the kidney.

9. Blood supply to the nephron is provided by the afferent arterioles, peritubular capillaries, and vasa recta.

**Sections 15.4, 15.5, 15.6**

10. The formation of urine involves three processes: glomerular filtration, tubular reabsorption, and tubular secretion.

11. The glomerular filtrate is produced by high blood pressure in the glomerular capillaries, and it contains water and small solutes in the same concentration as blood plasma.

12. Tubular reabsorption returns filtered water, sodium, major nutrients, and bicarbonate to the blood.

13. Tubular secretion moves substances from the capillaries into the tubules.

14. The kidneys can excrete excess water by producing dilute urine and can conserve water by producing concentrated urine.

**Sections 15.7, 15.8**

15. The kidneys contribute to homeostasis by adjusting blood volume and pressure, regulating salt balance, maintaining blood pH, controlling production of red blood cells, and participating in the formation of vitamin D.
16. Antidiuretic hormone influences water balance by increasing the water permeability of the collecting duct.
17. Blood volume is influenced by aldosterone, renin, and atrial natriuretic hormone.
18. The kidneys maintain blood pH by secreting $H^+$.
19. The kidneys stimulate red blood cell production by secreting erythropoietin.
20. Disorders of the urinary system include kidney stones, acute and chronic renal failure, and urinary tract infections.

# Exercises

Complete the exercises for each section after you have read and studied the section. If you cannot answer some questions, or answer them incorrectly, return to the chapter and review this information. You may find it helpful to work on only one section at a time. When you have completed all sections, take the Chapter Test as an indicator of your mastery of this topic.

15.1 **The urinary system contributes to homeostasis**

15.2 **Organs of the urinary system**

15.3 **Nephrons produce urine**

## Matching

_____ 1. **excretion**        a. the portion of a nephron that surrounds and encloses the glomerulus

_____ 2. **urinary system**   b. a muscular tube that transports urine to the bladder

_____ 3. **urea**             c. a vessel that carries blood away from the glomerulus

_____ 4. **kidneys**          d. blood vessels that supply the loop of Henle and collecting duct

_____ 5. **ureter**           e. portion of the nephron that is affected by aldosterone

_____ 6. **urinary bladder**  f. a vessel that carries blood into the nephron

_____ 7. **urethra**    g. processes that remove wastes and excess materials from the body

_____ 8. **nephron**    h. a tube that receives filtrate from many nephrons

_____ 9. **Bowman's capsule**    i. a capillary network formed by the efferent arterioles that carries away reabsorbed water, ions, and nutrients

_____ 10. **proximal tubule**    j. a network of capillaries that produces the filtrate

_____ 11. **loop of Henle**    k. the primary organs of the urinary system

_____ 12. **distal tubule**    l. the site of the most tubular reabsorption

_____ 13. **collecting duct**    m. the major nitrogenous waste product in urine

_____ 14. **afferent arteriole**    n. a muscular tube that carries urine from the body

_____ 15. **glomerulus**    o. the portion of the nephron that extends deep into the medulla

_____ 16. **efferent arterioles**    p. a body system that produces, transports, stores, and excretes urine

_____ 17. **peritubular capillaries**    q. the functional unit of the kidney

_____ 18. **vasa recta**    r. an organ that stores urine

## Labeling

*Use the terms below to label Figure 15.1.*

| kidney | ureter | renal pelvis |
| medulla | cortex | renal blood vessels |
| ureter | bladder | major blood vessels |
| nephron | urethra | |

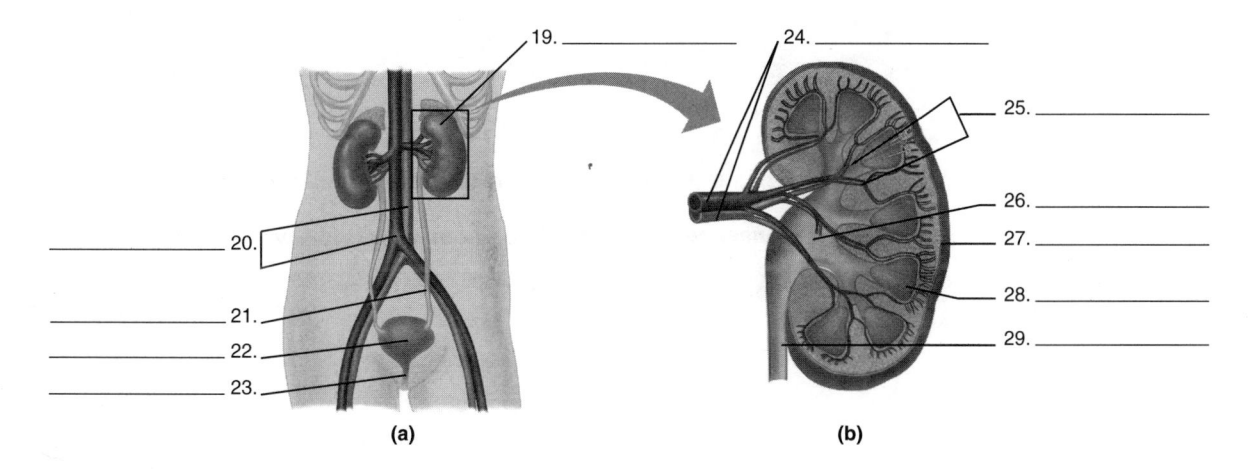

19. _____
20. _____
21. _____
22. _____
23. _____
24. _____
25. _____
26. _____
27. _____
28. _____
29. _____

(a)    (b)

**Figure 15.1**

## Labeling

*Use the terms below to label Figure 15.2.*

| glomerulus | loop of Henle | peritubular capillaries |
| artery | distal tubule | afferent arteriole |
| glomerular capsule | vein | collecting duct |
| ascending limb | proximal tubule | descending limb |
| efferent arteriole | | |

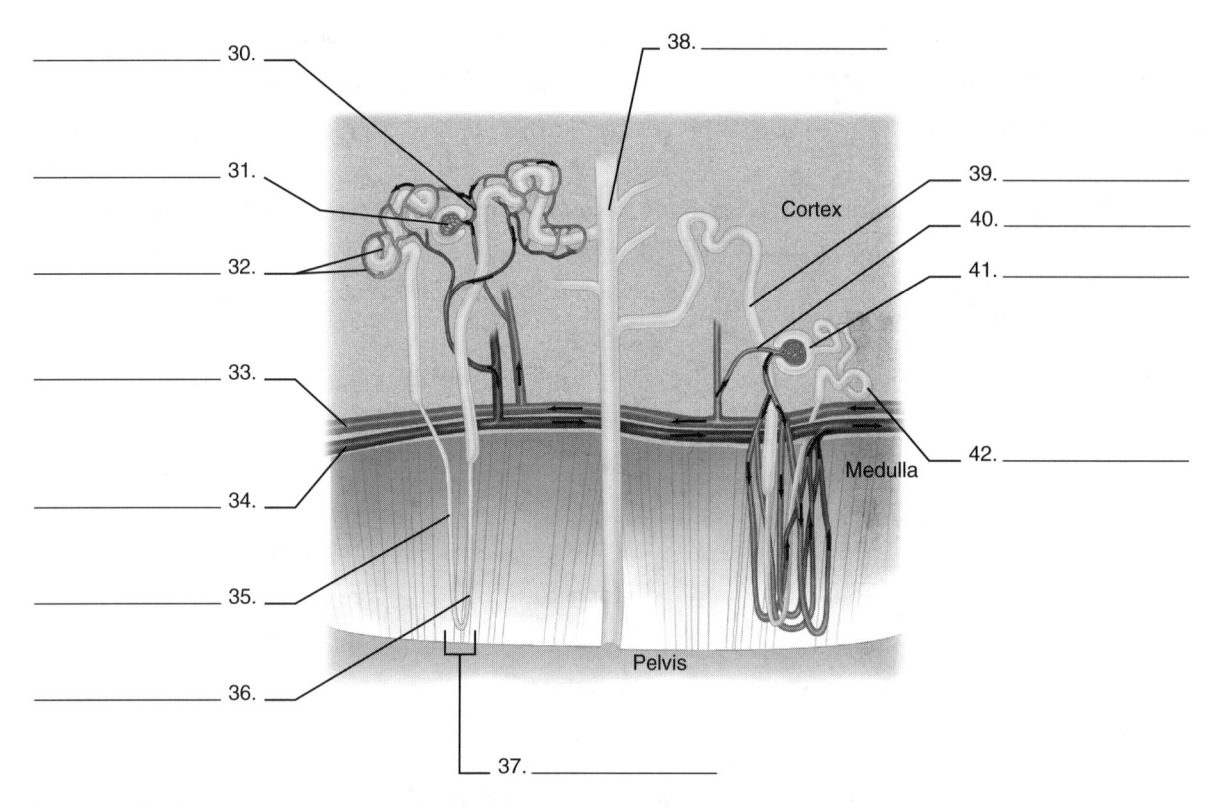

**Figure 15.2**

15.4 Formation of urine—filtration, reabsorption, and secretion

15.5 The kidneys can produce dilute or concentrated urine

15.6 Urination depends on a reflex

## Fill-in-the-Blank

*Referenced sections are in parentheses.*

1. During glomerular filtration, filtrate from the glomerulus moves into _____ _____. (15.4)

2. The peritubular capillaries add solutes to the fluid in the tubules during the process of _____ _____. (15.4)

3. Glomerular filtrate contains water and small _____, but not large _____ or blood _____. (15.4)

4. The driving force for glomerular filtration is high _____ _____ in the glomerular capillaries. (15.4)

5. Compared to most other capillaries, the capillaries of the glomerulus are more permeable to _____ and small _____, and less permeable to large _____. (15.4)

6. Most tubular reabsorption occurs in the _____ tubule. (15.4)

7. Substances that are almost completely reabsorbed include _____, _____, _____, and _____. (15.4)

8. The distal tubule and collecting duct reabsorb less than _____ of the water, but this is the region where water excretion is _____. (15.4)

9. Protein in the urine indicates possible glomerular damage and is called _____. (15.4)

10. Tubular secretion depends on the transport processes of _____ _____ and _____ _____. (15.4)

11. The proximal tubule secretes _____ and _____. (15.4)

12. The distal tubule secretes _____.

13. A dilute urine is produced when the kidneys reabsorb _____ water. (15.5)

14. The ability to form dilute urine depends partly on the _____ of solutes in the _____ _____. (15.5)

15. The solute concentration of the renal medulla is kept high by the recycling of _____. (15.5)

16. A concentrated urine is produced when the kidneys _____ more water. (15.5)

17. The movement of fluid in opposite directions in the ascending and descending limbs of the loop of Henle is called _____ flow. (15.5)

18. ADH _____ the permeability of the _____ _____ to water. (15.5)

19. In the presence of ADH, more _____ is reabsorbed, producing a _____ urine. (15.5)

20. Urination depends on a neural reflex called the _____ _____. (15.6)

21. The internal urethral sphincter consists of _____ muscle, and the external urethral sphincter consists of _____ muscles. (15.6)

## Completion

*For each characteristic listed below indicate the region of the nephron where it occurs. Nephron regions are labeled a–f, and may be used more than once. Some characteristics also occur in more than one region.*

**Region**

a. proximal tubule
b. descending limb of the loop of Henle
c. first portion of the ascending limb of the loop of Henle
d. final portion of the ascending limb of the loop of Henle
e. distal tubule
f. collecting duct

**Characteristic**

____ 22. impermeable to water

____ 23. water permeability in this region is influenced by ADH

____ 24. NaCl is actively transported

____ 25. impermeable to NaCl

____ 26. sodium is actively transported and chlorine moves by diffusion

____ 27. water moves by simple diffusion

### 15.7 The kidneys maintain homeostasis in many ways

### 15.8 Disorders of the urinary system

## Word Choice

*Circle the term that correctly completes each sentence.*

1. ADH is secreted into the (blood/kidney tubules) by the posterior pituitary when blood solute concentration (decreases/increases).

2. In response to ADH, the permeability of the collecting duct to water (decreases/increases), resulting in a more (dilute/concentrated) urine.

3. When blood solute concentration decreases, the (hypothalamus/kidney) signals the posterior pituitary to (decrease/increase) ADH secretion.

4. Aldosterone is produced by the (hypothalamus/posterior pituitary/adrenal gland).

5. Aldosterone (decreases/increases) the reabsorption of (sodium/chloride/water) across the distal tubule and collecting duct.

6. A decrease in blood volume triggers a(n) (decrease/increase) in aldosterone secretion.

7. A decrease in blood pressure causes (decreased/increased) secretion of rennin by cells in the (afferent/efferent) arteriole.

8. The effect of rennin on (angiotensinogen/angiotensin II) is the production of angiotensin I.

9. Angiotensin II stimulates the secretion of (ADH/aldosterone).

10. ANH is a hormone secreted by the (hypothalamus/heart/kidney) that causes (decreased/increased) Na$^+$ excretion in the urine.

## Paragraph Completion

*Use the following terms to complete the paragraphs, then repeat this exercise with the terms covered. Terms used more than once are listed multiple times.*

| H⁺ | decrease | vitamin D | excreted |
| kidneys | liver | ammonia | H⁺ |
| increase | bicarbonate | calcium | decreases |
| phosphorus | UV rays | erythropoietin | |

The kidneys help to maintain homeostasis in the body in many ways. In addition to monitoring blood volume and blood pressure, they influence acid-base balance, red blood cell production, and vitamin D synthesis. The kidneys help to maintain acid-base balance by secreting (11) _____ into the tubule, and by producing (12) _____. The hydrogen ions aid the reabsorption of (13) _____, and ammonia combines with excess (14) _____ and is (15) _____.

(16) _____ is a hormone produced by the kidneys when oxygen levels in the kidney (17) _____. Erythropoietin stimulates the bone marrow to (18) _____ red blood cell production. As oxygen levels return to normal, erythropoietin production (19) _____.

The kidneys also participate in (20) _____ synthesis. Vitamin D is necessary for the absorption of (21) _____ and (22) _____ from the digestive tract. Vitamin D synthesis begins in the skin when exposure to (23) _____ initiates the production of an inactive form of vitamin D. This inactive vitamin D precursor is transported first to the (24) _____ and then to the (25) _____ where it is converted to a useable form.

## Short Answer

26. What causes kidney stones and how are they treated?

27. What is the most common cause of urinary tract infections? Why are UTIs more common in women than in men?

28. How is acute renal failure different from chronic renal failure?

29. What are the benefits and risks of CAPD?

# Chapter Test

## Multiple Choice

1. Which of the following is not part of the urinary system?
   a. kidneys
   b. liver
   c. bladder
   d. ureters

2. The major nitrogenous waste product in urine is:
   a. urea.
   b. creatinine.
   c. bicarbonate.
   d. salt.

3. The functional unit of the kidney is the:
   a. renal medulla.
   b. pyramid.
   c. nephron.
   d. glomerulus.

4. The structure of the glomerular capillaries contributes to the formation of filtrate because:
   a. the walls of the capillaries are strengthened by smooth muscle to withstand the high blood pressure in the glomerulus.
   b. the walls of the capillaries contain tight junctions, preventing excess plasma from leaking into the glomerulus.
   c. the walls of the capillaries are much more permeable to water and small solutes than other capillaries.
   d. the branching of the capillaries provides for even distribution of blood pressure.

5. Which of the following incorrectly describes the process of urine formation?
   a. glomerular filtration—fluid filtered from the blood in the glomerulus moves into Bowman's capsule
   b. tubular reabsorption—fluid filtered from the proximal tubule is absorbed by the distal tubule
   c. tubular secretion—essential solutes are secreted from the tubules into the peritubularcapillaries
   d. b and c are incorrect
   e. a, b, and c are incorrect
   f. a, b, and c are all correctly described

6. Which of the following correctly summarizes how the process of urine formation contributes to the final urine content?
    a. glomerular filtration − tubular reabsorption + tubular secretion = excretion
    b. glomerular filtration + tubular secretion + tubular reabsorption = excretion
    c. tubular reabsorption + glomerular filtration − tubular secretion = excretion
    d. tubular secretion − glomerular filtration + tubular reabsorption = excretion

7. Blood pressure is the driving force that moves fluid out of all the capillaries in the body. Why is so much more fluid moved out of the capillaries in the glomerulus of the kidneys than in other areas of the body?
    a. Movement of fluid out of the glomerular capillaries is not opposed by osmotic pressure.
    b. Blood entering the kidneys contains more water than capillary blood in other parts of the body.
    c. The blood pressure in the glomerular capillaries is twice as high as in other capillaries of the body.
    d. Fluid is actively transported out of the glomerular capillaries.

8. The _____ arteriole carries blood into the glomerulus, and the _____ arteriole carries blood away.
    a. proximal, distal
    b. ascending, descending
    c. filtering, collecting
    d. afferent, efferent

9. Substances that are reabsorbed from the proximal tubule will enter the:
    a. afferent arteriole.
    b. peritubular capillaries.
    c. collecting duct.
    d. descending arteriole.

10. Most tubular reabsorption occurs in the _____ where lumen cells contain _____.
    a. proximal tubule, microvilli
    b. loop of Henle, active transport proteins
    c. collecting duct, pores
    d. renal medulla, proton pumps

11. Which of the following is not almost completely reabsorbed from the filtrate?
    a. glucose
    b. water
    c. creatinine
    d. salt

12. Sodium is reabsorbed from the proximal tubule by:
    a. passive diffusion.
    b. facilitated diffusion.
    c. active transport.
    d. osmosis.

13. What two ions are secreted by the proximal tubule in maintaining acid-base balance?
    a. hydrogen and bicarbonate
    b. urea and ammonium
    c. hydrogen and ammonium
    d. potassium and hydrogen

14. The diffusion of water out of the descending limb creates a gradient that favors:
    a. the diffusion of water into the ascending limb.
    b. the diffusion of NaCl out of the ascending limb.
    c. the active transport of glucose out of the ascending limb.
    d. the active transport of $H^+$ into the descending limb.

15. The action of ADH on the collecting duct produces:
    a. a dilute urine.
    b. a concentrated urine.
    c. urine with a high glucose content.
    d. the presence of protein in the urine.

16. Which urethral sphincter is voluntary?
    a. internal
    b. external
    c. secondary
    d. primary

17. The juxtaglomerular apparatus consists of cells of the:
    a. afferent arteriole and distal tubule.
    b. afferent arteriole and proximal tubule.
    c. efferent arteriole and distal tubule.
    d. efferent arteriole and collecting duct.

18. Caffeine acts as a mild diuretic because it:
    a. stimulates the release of ADH.
    b. stimulates sodium reabsorption.
    c. inhibits sodium reabsorption.
    d. increases the permeability of the glomerular capillaries.

19. The effect of aldosterone on the distal tubule and collecting duct is to:
    a. increase reabsorption of sodium.
    b. decrease reabsorption of sodium.
    c. increase excretion of water.
    d. increase secretion of $H^+$.

20. The greatest stimulus for release of aldosterone is:
    a. high sodium concentrations in the filtrate.
    b. decreased blood volume.
    c. increased blood volume.
    d. increased release of renin.

21. Renin is produced by:
    a. afferent arteriole cells in the juxtaglomerular apparatus.
    b. the liver.
    c. the pituitary.
    d. efferent arteriole cells in the juxtaglomerular apparatus.

22. The kidneys regulate the production of red blood cells by secreting:
    a. renin.
    b. erythropoietin.
    c. angiotensin I.
    d. vitamin D.

23. If the liver were unable to produce angiotensinogen, which of the following might occur?
    a. the kidneys would be unable to produce renin
    b. ADH levels would increase
    c. blood vessels would dilate
    d. angiotensin II would not be available

24. What is the role of the kidneys in the formation of vitamin D?
    a. the kidneys function alone in the synthesis of vitamin D
    b. the kidneys function with the skin and the liver in vitamin D synthesis
    c. the kidneys increase vitamin D secretion when blood pressure is low
    d. the kidneys secrete vitamin D and erythropoietin together

25. Which of the following are characteristics of acute renal failure?
    a. short-term duration
    b. incurable
    c. end-stage renal failure
    d. associated with Type I diabetes

# Key Concept Review Questions

*Each of the Key Concepts listed at the beginning of this chapter has been rewritten as a question below. After successfully completing the study guide exercises and the Chapter Test, you should be able to answer each of these questions. Refer to the Key Concepts list at the beginning of this chapter to check your answers.*

1. What is excretion?
2. What systems and organs contribute to excretion?
3. What are the organs of the urinary system? What are the functions of the urinary system?
4. How does the urinary system contribute to homeostasis?
5. What are the main organs of the urinary system?
6. What are the three main regions of the kidney?
7. How do the ureters, the bladder, and the urethra contribute to the function of the urinary system?
8. What is the functional unit of the kidney?
9. What vessels supply blood to a nephron?
10. What are the three processes involved in the formation of urine?
11. What pressure contributes to the formation of glomerular filtrate, and what does the filtrate contain?
12. What occurs in tubular reabsorption?
13. What occurs in tubular secretion?
14. What changes in water excretion do the kidneys make to produce dilute or concentrated urine?
15. What are five ways in which the kidneys contribute to homeostasis?
16. How does ADH influence water balance?
17. What are three hormones or chemicals that influence blood volume?
18. How do the kidneys maintain blood pH?
19. How do the kidneys stimulate red blood cell production?
20. What four disorders of the urinary system are discussed in this chapter?

# Answer Key

## Sections 15.1, 15.2, 15.3

**1.**g; **2.**p; **3.**m; **4.**k; **5.**b; **6.**r; **7.**n; **8.**q; **9.**a; **10.**l; **11.**o; **12.**e; **13.**h; **14.**f; **15.**j; **16.**c; **17.**i; **18.**d; **19.** kidney; **20.** major blood vessels; **21.** ureter; **22.** bladder; **23.** urethra; **24.** renal blood vessels; **25.** nephron; **26.** renal pelvis; **27.** cortex; **28.** medulla; **29.** ureter; **30.** efferent arterioles; **31.** glomerulus; **32.** peritubular capillaries; **33.** artery; **34.** vein; **35.** descending limb; **36.** ascending limb; **37.** loop of Henle; **38.** collecting duct; **39.** distal tubule; **40.** afferent arteriole; **41.** glomerular capsule; **42.** proximal tubule

## Sections 15.4, 15.5, 15.6

**1.** Bowman's capsule; **2.** tubular secretion; **3.** solutes, proteins, cells; **4.** blood pressure; **5.** water, solutes, protein; **6.** proximal; **7.** glucose, bicarbonate, water, sodium; **8.** 10%, regulated; **9.** proteinuria; **10.** active transport, facilitated diffusion; **11.** $H^+$, $NH_4^+$; **12.** $K^+$; **13.** less; **14.** concentration, renal medulla; **15.** urea; **16.** reabsorb; **17.** countercurrent; **18.** increases, collecting duct; **19.** water, concentrated; **20.** micturition reflex; **21.** smooth, skeletal; **22.**c,d; **23.**f; **24.**d,e,f; **25.**b; **26.**a; **27.**a,b,f

## Sections 15.7, 15.8

**1.** blood, increases; **2.** increases, concentrated; **3.** hypothalamus, decrease; **4.** adrenal gland; **5.** increases, sodium; **6.** increase; **7.** increased, afferent; **8.** angiotensinogen; **9.** aldosterone; **10.** heart, increased; **11.**$H^+$; **12.** ammonia; **13.** bicarbonate; **14.** $H^+$; **15.** excreted; **16.** Erythropoietin; **17.** decrease; **18.** increase; **19.** decreases; **20.** vitamin D; **21.** calcium; **22.** phosphorus; **23.** UV rays; **24.** liver; **25.** kidneys; **26.** crystallization of minerals in the urine, they often require no treatment but can be removed surgically or crushed when necessary; **27.** bacteria, because the urethra is shorter in women; **28.** acute renal failure is short-term and often correctable, chronic renal failure is long-term and irreversible; **29.** benefits include at-home use and freedom of movement, infection is a risk

## Chapter Test

**1.**b; **2.**a; **3.**c; **4.**c; **5.**d; **6.**a; **7.**c; **8.**d; **9.**b; **10.**a; **11.**c; **12.**c; **13.**c; **14.**b; **15.**b; **16.**b; **17.**a; **18.**c; **19.**a; **20.**b; **21.**a; **22.**b; **23.**d; **24.**b; **25.**a

# 16

# Reproductive Systems

## Chapter Summary and Key Concepts

*After reading and studying this chapter you should know the following:*

### Section 16.1

1. The reproductive system consists of primary reproductive organs, accessory glands, and ducts.

2. The primary organs of the male reproductive system are the testes, which secrete testosterone and produce sperm.

3. Sperm are produced in the seminiferous tubules and mature in the epididymis.

4. Sperm leave the body through the vas deferens and urethra in a fluid called semen.

5. Semen contains sperm and glandular secretions that provide nutrients, alkaline fluid, and mucus.

6. Sperm cells contain 23 chromosomes and are produced by a cell division process called meiosis.

7. Testosterone, LH, and FSH regulate the male reproductive system.

### Section 16.2

8. The primary organs of the female reproductive system are the ovaries, which release oocytes.

9. The oviduct is a hollow tube that transports the oocyte from the ovary to the uterus, and it is the site of fertilization.

10. The uterus supports the development of a fertilized egg, and it consists of an inner endometrium and an outer myometrium.

11. The vagina is the female organ of intercourse and the birth canal.

12. The mammary glands are accessory organs that produce and store milk.

### Sections 16.3, 16.4, 16.5

13. The ovaries and uterus undergo a monthly pattern of changes called the menstrual cycle, which consists of an ovarian cycle and a uterine cycle.

14. The ovarian cycle results in the release of a secondary oocyte capable of being fertilized.

15. The uterine cycle prepares the endometrium of the uterus to accept and nurture a fertilized egg.

16. GnRH, LH, FSH, estrogen, and progesterone are hormones involved in regulating the female reproductive system.

17. The human sexual response is a series of events that coordinate sexual function to accomplish intercourse and fertilization.

18. Birth control methods are designed to prevent fertilization.

19. Disorders of the reproductive system include infertility, complications of pregnancy, cancers and tumors of the reproductive organs, and sexually transmitted diseases.

20. STDs caused by bacteria are curable, while those caused by viruses are not. STDs include gonorrhea, syphilis, chlamydia, AIDS, hepatitis B, herpes, and genital warts.

# Exercises

*Complete the exercises for each section after you have read and studied the section. If you cannot answer some questions, or answer them incorrectly, return to the chapter and review this information. You may find it helpful to work on only one section at a time. When you have completed all sections, take the Chapter Test as an indicator of your mastery of this topic.*

# Labeling

*Label each of the structures indicated in Figure 16.1.*

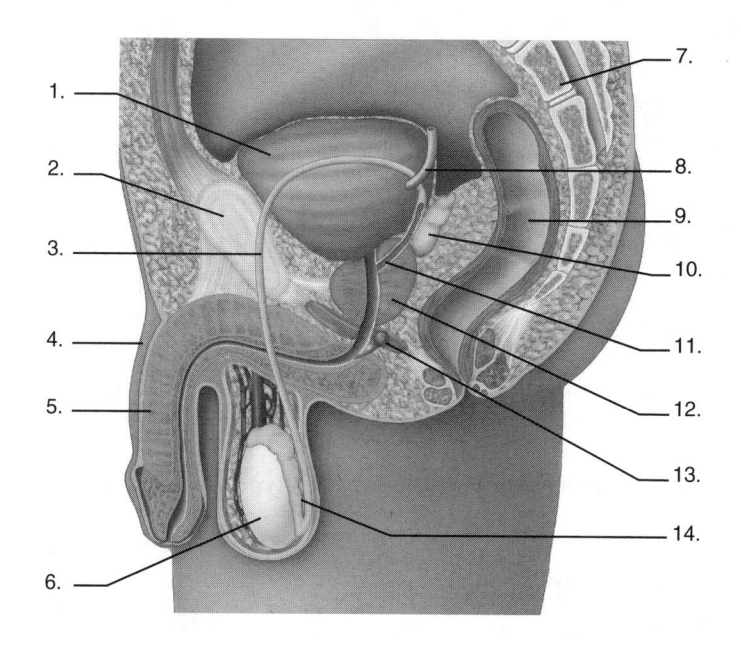

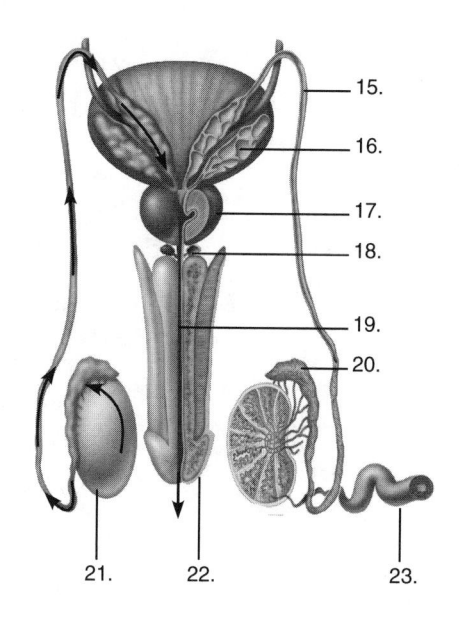

**Figure 16.1**

1. _____
2. _____
3. _____
4. _____
5. _____
6. _____
7. _____
8. _____
9. _____
10. _____
11. _____
12. _____
13. _____
14. _____
15. _____
16. _____
17. _____
18. _____
19. _____
20. _____
21. _____
22. _____
23. _____

### 16.1 Male reproductive system delivers sperm

## Fill-in-the-Blank

24. Sperm are the male _____.
25. In humans, diploid cells have _____ chromosomes and haploid cells have _____ chromosomes.
26. Secretion of FSH by the anterior pituitary is inhibited by the hormone _____.
27. Blood levels of testosterone are maintained by a _____ feedback loop involving the hypothalamus and the _____ _____.
28. Decreased levels of testosterone stimulate the hypothalamus to secrete _____.
29. _____ released by the pituitary increases secretion of testosterone.
30. When the temperature of developing sperm is too low, the _____ brings the testes nearer to the body.
31. Secretions from the _____ _____ may help to induce muscle contraction in the female reproductive tract.
32. Sperm leave the male body in a fluid called _____.
33. Secretions from the _____ gland provide lubrication for sexual intercourse.
34. The seminiferous tubules contain _____ cells that nourish developing sperm.

## Completion

*Indicate the correct structure or function to complete the table below.*

| Structure | Function |
|---|---|
| 35. | production of sperm and testosterone |
| Scrotum | 36. |
| 37. | location within the testes where sperm are produced |
| Epididymis | 38. |
| 39. | transports sperm from the epididymis |
| 40. | transports sperm from the vas deferens to the urethra |
| Urethra | 41. |
| Seminal vesicle | 42. |
| 43. | gland that adds an alkaline fluid to semen |
| 44. | gland that secretes mucus into the urethra |

## Labeling

*In Figure 16.2a, label each type of cell indicated. In Figure 16.2b, label each region of a mature sperm. Then answer the questions that follow.*

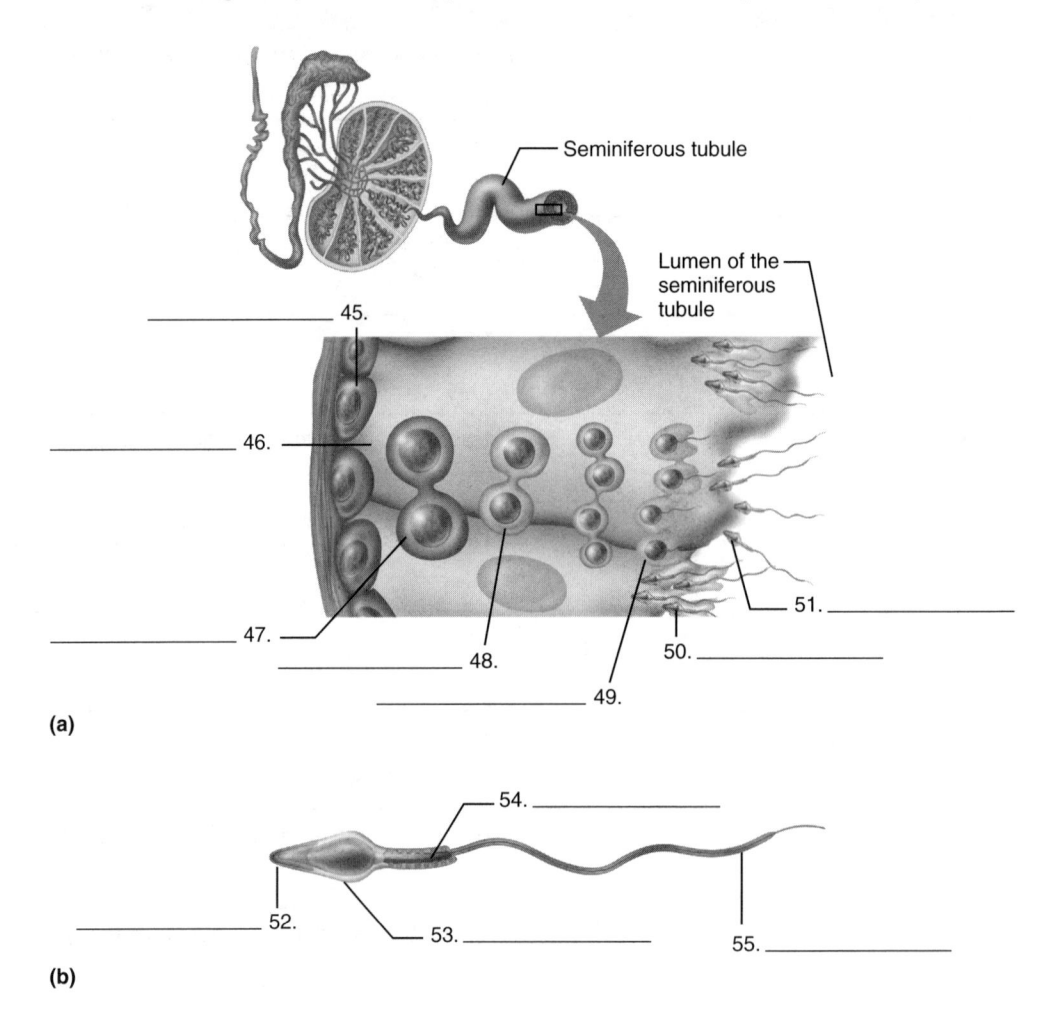

**Figure 16.2**

56. Are spermatogonia haploid or diploid?
57. Are sperm haploid or diploid?
58. What type of cell division produces primary spermatocytes?
59. What type of sperm cell is produced by Meiosis I?
60. What type of sperm cell is produced by Meiosis II?
61. Are spermatids capable of fertilizing an egg?
62. What is contained in the head of a mature sperm?
63. What is contained in the acrosome?
64. What is the function of the midpiece?
65. What is the function of the tail?

## 16.2 Female reproductive system produces eggs and supports pregnancy

### Matching

___ 1. **ovaries**  a. the female external genitalia
___ 2. **oocytes**  b. the production of milk by the mammary glands
___ 3. **estrogen and progesterone**  c. a small organ that contains erectile tissue
___ 4. **oviduct**  d. the birth canal
___ 5. **uterus**  e. female reproductive organs, site of storage and development of oocytes
___ 6. **endometrium**  f. modified sweat glands containing milk-producing lobules
___ 7. **cervix**  g. female sex hormones
___ 8. **vagina**  h. the outer skin folds of the vulva
___ 9. **vulva**  i. a hollow organ that supports the growth and development of a fertilized egg
___ 10. **labia majora**  j. female gametes
___ 11. **labia minora**  k. the inner layer of the uterus
___ 12. **clitoris**  l. a tube leading from the ovary to the uterus
___ 13. **mammary glands**  m. narrow end of the uterus, which leads into the vagina
___ 14. **lactation**  n. the smaller inner folds of the vulva

### Fill-in-the-Blank

15. Fertilization occurs in the _____.

16. In the process of implantation, a fertilized egg attaches to the _____.

17. The _____ is the pigmented area of the breast that surrounds the nipple.

18. _____ glands in the breast are specialized to produce milk.

19. The hormone _____ stimulates lactation, and the hormone _____ stimulates the release of milk.

## Labeling

*Use the terms below to label Figure 16.3.*

| clitoris | uterus | vagina | labium minora |
| oviduct | ovary | fimbriae | bladder |
| labium majora | urethra | cervix | |

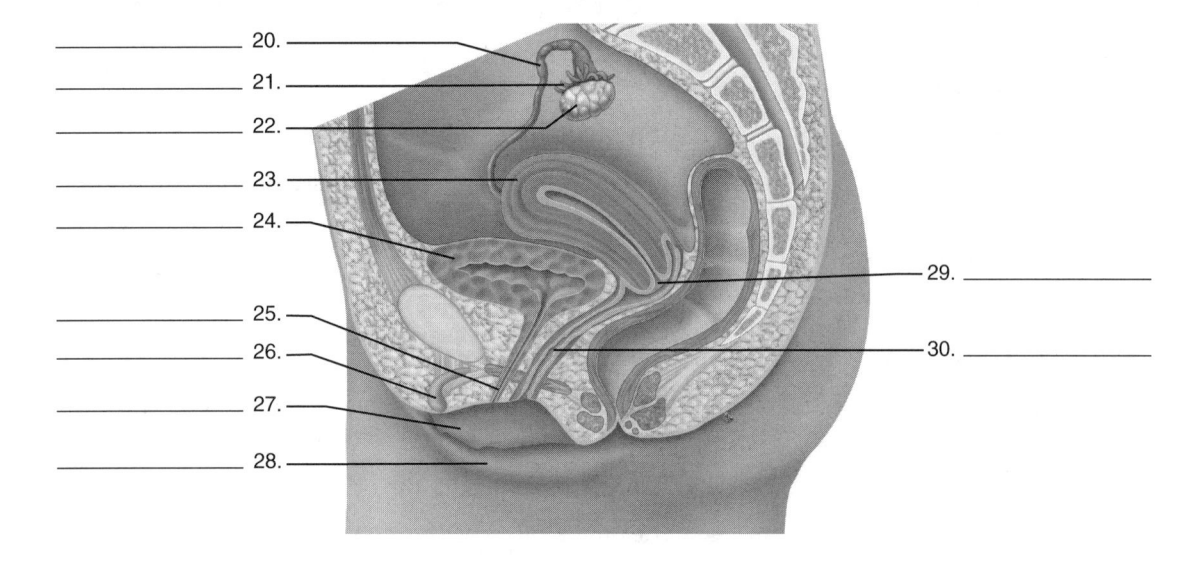

20. _____
21. _____
22. _____
23. _____
24. _____
25. _____
26. _____
27. _____
28. _____
29. _____
30. _____

**Figure 16.3**

**16.3** The menstrual cycle consists of ovarian and uterine cycles

**16.4** Human sexual response, intercourse, and fertilization

**16.5** Birth control methods: Controlling infertility

## Completion

*Use the terms below to complete the paragraph, then repeat this exercise with the terms covered.*

| endometrium | secondary | pellucida | ovulation |
| ovarian | Meiosis II | corpus | Graafian |
| antrum | anterior | LH | puberty |
| FSH | ovarian | follicle | decrease |
| granulosa | primary | Meiosis I | zona |
| pituitary | ovulation | GnRH | hypothalamus |
| progesterone | luteum | glycoprotein | Meiosis I |

The (1) _____ cycle involves the maturation and release of oocytes from the ovary. Each ovary contains approximately one million (2) _____ oocytes at birth. The oocytes have already begun the process of (3) _____. Continued development of an oocyte does not begin until (4) _____, and an oocyte will not complete (5) _____ until a sperm enters.

Each primary oocyte is surrounded by nourishing cells called (6) _____ cells. These cells, along with the oocyte, make up an immature (7) _____. About a dozen immature follicles begin the development process each month in response to (8) _____ and LH released by the (9) _____ (10) _____. At the beginning of the ovarian cycle, (11) _____ from the (12) _____ increases the concentration of these hormones, triggering oocyte development. Usually only one oocyte will complete the maturation process and be released during (13) _____.

As an immature follicle grows, the granulosa cells begin to secrete (14) _____ that become a noncellular coating called the (15) _____ (16) _____. Next, a fluid-filled space called the (17) _____ develops, and the follicle begins secreting estrogen and (18) _____. The primary ooctye will now finish the process of (19) _____, producing a (20) _____ oocyte and a polar body. This mature follicle is now called a (21) _____ follicle.

As estrogen (produced by the follicle) levels rise, the pituitary is stimulated to release a surge of the hormone (22) _____. This hormone surge triggers the process of (23) _____, the follicle ruptures, and the secondary oocyte is released. The remnants of the follicle will now form the (24) _____ (25) _____, which secretes large amounts of estrogen and progesterone to prepare the (26) _____ for implantation, if fertilization should occur. If fertilization does not occur, the corpus luteum degenerates, estrogen and progesterone levels (27) _____, and the (28) _____ cycle will begin again.

## Short Answer

*Refer to Figure 16.4 of the menstrual cycle to answer the questions.*

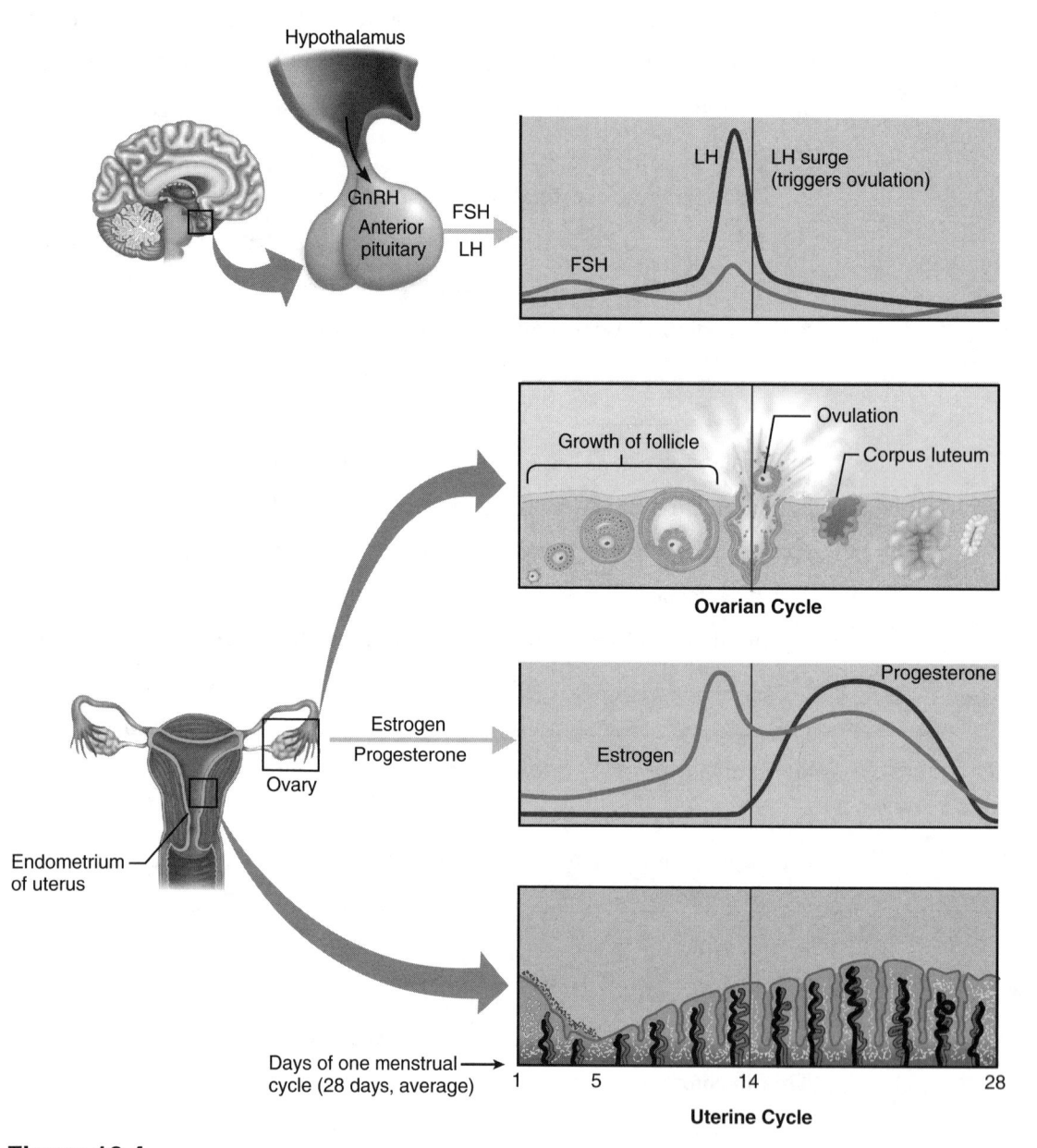

**Figure 16.4**

29. What days constitute the menstrual phase?

30. What body levels of estrogen and progesterone contribute to the events of the menstrual phase?

31. What days constitute the proliferative phase?

32. What hormone activity triggers the proliferative phase?

33. What event of the ovarian cycle coincides with the end of the proliferative phase?

34. What uterine phase covers the last half of the menstrual cycle?

35. What hormone level is the highest during the last half of the menstrual cycle?

36. What event is the uterus preparing for during this phase?

37. In the absence of fertilization, what event in the ovarian cycle causes the levels of estrogen and progesterone to decline?

38. Does positive feedback influence the menstrual cycle before or after ovulation?

39. Describe this positive feedback loop.

40. What prevents development of a second follicle during the luteal phase?

## Completion

*Name each birth control method described in the table.*

| Method of Birth Control | Description |
|---|---|
| 41. | the vas deferens is tied in two places and the segment between the ties is removed |
| 42. | consists of an injection of progesterone that lasts 3 months, inhibiting FSH and LH |
| 43. | a chemical in the form of a foam, cream, jelly, or douche that destroys sperm |
| 44. | a plastic or metal piece inserted into the uterus that causes inflammation, interfering with implantation |
| 45. | intercourse is avoided from about 5 days before ovulation until 3 days after ovulation |
| 46. | a monthly injection of estrogen and progesterone |
| 47. | a patch that slowly releases hormones into the bloodstream |
| 48. | blocks the action of progesterone, causing a chemical abortion |

**16.6 Infertility: Inability to conceive**

**16.7 Sexually transmitted diseases: A worldwide concern**

## Short Answer

1. What is the definition of infertility?

2. What is the minimum sperm count required for male fertility?

3. What is the most common cause of female infertility?

4. Briefly describe the process of invitro fertilization (IVF).

5. How does gamete intrafallopian transfer (GIFT) differ from IVF?

6. What is the most prevalent STD?

## Completion

*For each complication or disease of the reproductive system listed below, indicate the cause, symptoms, and treatment.*

7. Pelvic inflammatory disease

    a. cause:

    b. symptoms:

    c. treatment:

8. Endometriosis

    a. cause:

    b. symptoms:

    c. treatment:

9. Gonorrhea
   a. cause:

   b. symptoms:

   c. treatment:

10. Syphilis
    a. cause:

    b. symptoms:

    c. treatment:

11. Chlamydia
    a. cause:

    b. symptoms:

    c. treatment:

12. Genital herpes
    a. cause:

    b. symptoms:

    c. treatment:

13. Genital warts
    a. cause:

    b. symptoms:

    c. treatment:

# Chapter Test

## Multiple Choice

1. Which of the following is a secondary sexual characteristic?
   a. testosterone
   b. testis
   c. seminal vesicle
   d. muscle mass

2. The male gamete is the _____, and the female gamete is the _____.
   a. testis, ovary
   b. testosterone, estrogen
   c. vas deferens, oviduct
   d. sperm, egg

3. Which of the following is not a characteristic of Sertoli cells?
   a. located between the seminiferous tubules
   b. secrete inhibin
   c. stimulated by testosterone
   d. support developing sperm

4. The epididymis:
   a. is a tube connecting the vas deferens to the urethra.
   b. delivers secretions of the seminal vesicles into the urethra.
   c. is a coiled duct just outside the testis.
   d. is a mass of tightly coiled tubules where sperm are produced.

5. From production to exiting the male body, the pathway of a sperm will be:
   a. epididymis, seminiferous tubules, vas deferens, ejaculatory duct, urethra.
   b. vas deferens, ejaculatory duct, seminiferous tubules, epididymis, vas deferens.
   c. seminiferous tubules, vas deferens, epididymis, urethra, ejaculatory duct.
   d. seminiferous tubules, epididymis, vas deferens, ejaculatory duct, urethra.

6. In the process of sperm development, secondary spermatocytes develop into _____ which mature to become _____.
   a. primary spermatocytes, spermatids
   b. primary spermatocytes, sperm
   c. spermatids, sperm
   d. spermatids, spermatogonia

7. Which region of the sperm is incorrectly described?
   a. midpiece—carries secretions from the male glands that help to neutralize acidity in the vagina
   b. head—contains the male chromosomes
   c. acrosome—contains enzymes to aid penetration of the egg
   d. tail—provides for movement of the sperm through the female reproductive tract

8. Spermatogonia are _____ and sperm are _____.
   a. haploid, haploid
   b. haploid, diploid
   c. diploid, haploid
   d. diploid, diploid

9. Sperm production:
   a. is controlled by a positive feedback loop.
   b. occurs in the seminiferous tubules.
   c. is stimulated by inhibin.
   d. begins shortly after birth and continues through old age.

10. The secretion of testosterone is initiated by _____ released by the hypothalamus.
    a. GnRH
    b. LH
    c. FSH
    d. ADH

11. If the pathway from the uterus to the oviduct is blocked:
    a. an oocyte cannot be released.
    b. an oocyte cannot enter the oviduct.
    c. fertilization cannot occur.
    d. the uterine cycle will be disrupted.

12. The _____ provides the muscular force to expel a mature fetus during labor and birth.
    a. fimbriae
    b. endometrium
    c. myometrium
    d. cervix

13. At birth, the ovary:
    a. already contains approximately one million secondary oocytes.
    b. produces high levels of LH to prevent maturation of follicles until puberty.
    c. is controlled by inhibin released by the anterior pituitary.
    d. contains many more immature follicles than will be present at puberty.

14. Which of the following is not released from the ovary with the secondary oocyte?
    a. the polar body
    b. the zona pellucida
    c. the granulosa cells
    d. the corpus luteum

15. The function of the corpus luteum is to:
    a. protect the secondary oocyte from fertilization by multiple sperm.
    b. nourish the secondary ooctye until it reaches the uterus.
    c. secrete progesterone and estrogen to prepare the uterus for implantation.
    d. secrete progesterone and estrogen to prevent release of a second egg before the end of the uterine cycle.

16. Menstruation is triggered by:
    a. the implantation of a fertilized egg.
    b. the process of ovulation.
    c. a decline in the level of progesterone and estrogen.
    d. an increase in the level of progesterone and estrogen.

17. If fertilization does not occur, changing hormone levels leading to menstruation will result from:
    a. the absence of a chorion.
    b. the degeneration of the corpus luteum.
    c. the growth of a new follicle.
    d. proliferation of the endometrium.

18. The first half of the menstrual cycle is influenced by positive feedback. This occurs when:
    a. rising levels of estrogen in the first half of the cycle trigger a surge in LH.
    b. high levels of estrogen and progesterone inhibit LH and FSH.
    c. decreasing levels of estrogen stimulate an increase in LH.
    d. decreasing levels of progesterone stimulate an increase in LH and FSH.

19. Erection occurs when:
    a. arterioles leading into the vascular compartments of the penis dilate, while veins draining the penis are compressed.
    b. arterioles leading into the vascular compartment of the penis constrict, while veins draining the penis dilate.
    c. muscular contractions of the ejaculatory duct compress the vessels that supply the vascular compartments of the penis.
    d. smooth muscle in the penis contracts due to stimulation by the sympathetic nervous system.

20. Hormonal methods of birth control administer progesterone and estrogen in amounts that:
    a. stimulate the release of GnRH.
    b. stimulate the release of LH and FSH.
    c. inhibit the release of LH and FSH.
    d. prevent the proliferative phase of the uterine cycle.

21. One of the drawbacks in using a diaphragm is:
    a. the insertion of the diaphragm must be done in a doctor's office.
    b. a high failure rate.
    c. that diaphragms do not provide protection against STDs.
    d. the side effects, which may include headaches and nausea.

22. Male infertility usually results from:
    a. unusually high levels of testosterone.
    b. failure of the testis to descend completely into the scrotum.
    c. insufficient numbers of normal sperm.
    d. adrenal tumors resulting in increased estrogen secretion.

23. Endometriosis is a condition in which:
    a. the endometrium fails to proliferate during the uterine cycle.
    b. the endometrium is insensitive to estrogen and progesterone.
    c. there is abnormal thickening of the myometrium, creating pressure on the endometrium.
    d. endometrial tissue migrates up the oviduct and grows on other organs.

24. Which of the following types of STDs is incorrectly matched with its causative organism?
    a. gonorrhea—bacteria
    b. Hepatitis B—virus
    c. genital warts—virus
    d. syphilis—virus

25. Chlamydia:
    a. is the most common incurable STD.
    b. may have mild symptoms that go unnoticed.
    c. does not present a serious risk for PID.
    d. may result in sterility if not treated.
    e. b and d

# Key Concept Review Questions

*Each of the Key Concepts listed at the beginning of this chapter has been rewritten as a question below. After successfully completing the study guide exercises and the Chapter Test, you should be able to answer each of these questions. Refer to the Key Concepts list at the beginning of this chapter to check your answers.*

1. What are the general structures of a reproductive system?
2. What are the primary organs of the male reproductive system, and what are their functions?
3. Where are sperm produced, and where do they mature?
4. What tubular structures carry the sperm outside the body? What fluid contains the sperm?
5. What is contained in semen?
6. How many chromosomes are there in a sperm cell? What type of cell division produces sperm?
7. What are three hormones important in regulating the male reproductive system?
8. What are the primary organs of the female reproductive system, and what is their function?
9. What is the oviduct, and what is its function? Where does fertilization occur?
10. What is the role of the uterus? Name the two layers of the uterus.
11. What are two functions of the vagina?
12. What is the function of the mammary glands?
13. What are the two organs involved in the menstrual cycle?
14. Which cycle results in the release of a secondary oocyte?
15. Which cycle prepares the endometrium to accept and nurture a fertilized egg?
16. What are five hormones involved in regulating the female reproductive cycle?
17. The human sexual response coordinates sexual function to accomplish what two events?
18. Birth control methods are designed to prevent what general event?
19. What are the four general classes of reproductive disorders?
20. Which STDs are curable and which are not curable? List several STDs of each type.

# Answer Key

## Section 16.1

**1.** bladder; **2.** pubic symphysis; **3.** vas deferens; **4.** penis; **5.** erectile tissue; **6.** testis; **7.** vertebral column; **8.** ureter; **9.** rectum; **10.** seminal vesicle; **11.** ejaculatory duct; **12.** prostate gland; **13.** bulbourethral gland; **14.** epididymis; **15.** vas deferens; **16.** seminal vesicle; **17.** prostate gland; **18.** bulbourethral gland; **19.** urethra; **20.** epididymis; **21.** testis; **22.** penis; **23.** seminiferous tubule; **24.** gamete; **25.** 46, 23; **26.** inhibin; **27.** negative, pituitary gland; **28.** GnRH; **29.** LH; **30.** scrotum; **31.** seminal vesicles; **32.** semen; **33.** bulbourethral; **34.** Sertoli; **35.** testis; **36.** regulates temperature for developing sperm; **37.** seminiferous tubules; **38.** site of sperm maturation and storage; **39.** vas deferens; **40.** ejaculatory duct; **41.** transports sperm outside the male body; **42.** gland that produces a watery mixture of fructose and prostaglandins; **43.** prostate gland; **44.** bulbourethral gland; **45.** spermatogonium; **46.** Sertoli cell; **47.** primary spermatocyte; **48.** secondary spermatocyte; **49.** early spermatids; **50.** late spermatids; **51.** immature sperm; **52.** acrosome; **53.** head; **54.** midpiece; **55.** tail; **56.** diploid; **57.** haploid; **58.** mitosis; **59.** secondary spermatocyte; **60.** spermatid; **61.** no; **62.** DNA; **63.** enzymes that aid penetration of the egg; **64.** to house mitochondria that provide energy for sperm movement; **65.** movement of sperm

## Section 16.2

**1.**e; **2.**j; **3.**g; **4.**l; **5.**i; **6.**k; **7.**m; **8.**d; **9.**a; **10.**h; **11.**n; **12.**c; **13.**f; **14.**b; **15.** oviduct; **16.** endometrium; **17.** areola; **18.** Mammary; **19.** prolactin, oxytocin; **20.** oviduct; **21.** fimbriae; **22.** ovary; **23.** uterus; **24.** bladder; **25.** urethra; **26.** clitoris; **27.** labium minora; **28.** labium majora; **29.** cervix; **30.** vagina

## Sections 16.3, 16.4, 16.5

**1.** ovarian; **2.** primary; **3.** Meiosis I; **4.** puberty; **5.** Meiosis II; **6.** granulose; **7.** follicle; **8.** FSH; **9.** anterior; **10.** pituitary; **11.** GnRH; **12.** hypothalamus; **13.** ovulation; **14.** glycoprotein; **15.** zona; **16.** pellucida; **17.** antrum; **18.** progesterone; **19.** Meiosis I; **20.** secondary; **21.** Graafian; **22.** LH; **23.** ovulation; **24.** corpus; **25.** luteum; **26.** endometrium; **27.** decrease; **28.** ovarian; **29.** days 1–5; **30.** estrogen and progesterone levels decline; **31.** days 6–14; **32.** rising levels of estrogen and progesterone; **33.** ovulation; **34.** secretory phase; **35.** progesterone; **36.** implantation of a fertilized egg; **37.** degeneration of the corpus luteum; **38.** before; **39.** rising levels of estrogen in the first half of the proliferative phase trigger a surge in LH; **40.** high levels progesterone and estrogen, secreted by the corpus luteum, feed back to the hypothalamus and inhibit FSH secretion; **41.** vasectomy; **42.** Depo-Provera; **43.** spermicide; **44.** intrauterine device (IUD); **45.** rhythm method; **46.** Lunelle™; **47.** Ortho Evra®; **48.** Mifeprex®

## Sections 16.6, 16.7

**1.** failure to achieve pregnancy after one year of trying; **2.** 60 million sperm per ejaculate; **3.** PID: **4.** eggs are extracted from the female, fertilized outside the body, and embryos are implanted into the uterus; **5.** in GIFT fertilization occurs within the body when unfertilized eggs and sperm are surgically placed into a fallopian tube; **6.** Chlamydia; **7.** a. bacterial infection, b. infertility, irregular menstrual cycles; c. treatment is difficult; **8.** a. unknown, occurs when endometrial tissue grows outside the uterus, b. pain, infertility; c. surgery, drugs, hormone therapy; **9.** a. bacterial infection, b. symptoms include discharge and painful urination, infected individuals may have no symptoms, c. medication; **10.** a. bacterial infection, b. occurs in three stages, refer to text for symptoms; c. medication; **11.** a. bacterial infection, b. discharge, painful urination, swollen lymph nodes, PID, c. medication; **12.** a. viral infection, b. blisters, painful urination, swollen lymph nodes in groin, fever, c. no cure available, suppressive drugs only; **13.** a. viral infection, b. warts in genital area, c. no cure available, can be removed by surgery or freezing but tend to reappear

# Chapter Test

**1.**d; **2.**d; **3.**a; **4.**c; **5.**d; **6.**c; **7.**a; **8.**c; **9.**b; **10.**a; **11.**c; **12.**c; **13.**d; **14.**d; **15.**c; **16.**c; **17.**b; **18.**a; **19.**a; **20.**c; **21.**c; **22.**c; **23.**d; **24.**d; **25.**e

# 17

# Cell Reproduction and Differentiation

## Chapter Summary and Key Concepts

*After reading and studying this chapter you should know the following:*

**Sections 17.1, 17.2**

1. The cell cycle consists of two phases: interphase and the mitotic phase.

2. Interphase is a growth phase consisting of three stages: G1, S, and G2. During G1, cells grow; during S, the deoxyribonucleic acid (DNA) is replicated; and during G2, cells continue to grow in preparation for cell division.

3. The mitotic phase consists of a nuclear division called mitosis, and a cytoplasmic division called cytokinesis.

4. Human cells contain DNA arranged into 46 chromosomes; each chromosome includes segments of DNA, called genes, that encode a particular protein.

5. Replication is the process of copying a cell's DNA and occurs during the S phase of the cell cycle.

6. A replicated chromosome consists of two sister chromatids, joined by a centromere.

7. Transcription and translation occur when the information in a DNA gene is used to make a protein.

8. Transcription occurs in the nucleus and makes an RNA copy of a DNA gene.

9. Translation occurs in the cytoplasm and results in production of a protein.

10. Three types of RNA participate in translation: messenger RNA carries instructions about how to build the protein, transfer RNA transports amino acids to the growing protein, and ribosomal RNA forms the ribosome.

11. A mutation is a permanent change in the DNA.

12. Mutations may be helpful or harmful, and if not repaired, they may contribute to the process of evolution.

**Sections 17.3, 17.4, 17.5**

13. Mitosis can be divided into a series of phases called prophase, metaphase, anaphase, and telophase.
14. Mitosis produces two diploid daughter cells that are identical to the parent cell.
15. Meiosis consists of two successive nuclear divisions called meiosis I and meiosis II.
16. Meiosis produces four haploid cells that are genetically different from the parent cell.
17. In a female, meiosis results in the production of one functional egg. In a male, meiosis results in the production of four functional sperm.
18. Control of the cell cycle occurs through regulatory proteins called cyclins and the use of cell cycle checkpoints.
19. Environmental factors and selective gene expression lead to the development of specialized cells.
20. Differentiation is the process whereby cells become different from each other.

# Exercises

*Complete the exercises for each section after you have read and studied the section. If you cannot answer some questions, or answer them incorrectly, return to the chapter and review this information. You may find it helpful to work on only one section at a time. When you have completed all sections, take the Chapter Test as an indicator of your mastery of this topic.*

**17.1 The cell cycle creates new cells**

**17.2 Replication, transcription, and translation—an overview**

## Completion

*Use the letters in the list of characteristics below and in Figure 17.1 to complete each phase of the cell cycle in the table below. Some characteristics may be used more than once.*

**Characteristics**

a. includes mitosis and cytokinesis
b. the cell is at its smallest size
c. lasts 7–8 hours in mammalian cells
d. a phase in which healthy cells are not dividing
e. DNA is replicated
f. cell growth is occurring
g. DNA appears as chromatin
h. lasts 30–45 minutes in mammalian cells
i. sister chromatids are produced
j. includes all stages of the cell cycle that occur between cell divisions
k. the cell prepares for cell division

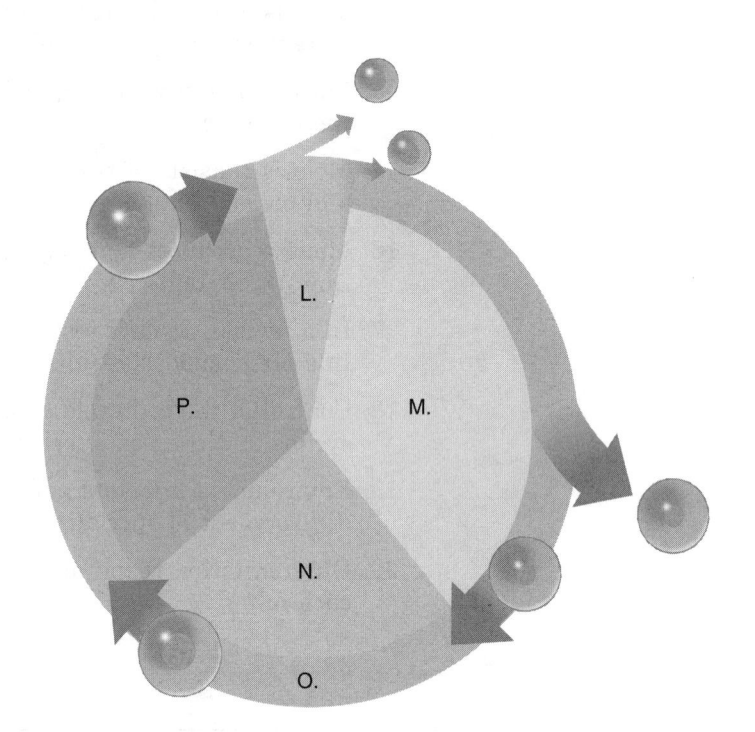

**Figure 17.1**

| Phase | Diagram Labeled Phase | Characteristics |
|---|---|---|
| $G_0$ | 1. not labeled | 2. |
| $G_1$ | 3. | 4. |
| S | 5. | 6. |
| $G_2$ | 7. | 8. |
| M | 9. | 10. |
| Interphase | 11. | 12. |

## Fill-in-the-Blank

13. The two periods of the cell cycle are the _____ and the _____ _____. (17.1)

14. The division of the nucleus occurs during the _____ _____. (17.1)

15. The division of the cytoplasm occurs during _____. (17.1)

16. Division of the cytoplasm produces two new _____ cells. (17.1)

17. Inside the nucleus, DNA is organized into separate structures called _____. (17.2)

18. Humans have _____ chromosomes in all cells that contain a nucleus except reproductive cells. (17.2)

19. Chromosomes contain DNA complexed with proteins called _____. (17.2)

20. During the S phase of the cell cycle, duplicated chromosomes are produced that consist of two identical _____ _____ joined by a _____. (17.2)

21. A _____ is the smallest identified functional unit of DNA. (17.2)

22. Chromosomes are found only in the _____ because they are too large to escape through nuclear _____. (17.2)

23. DNA is used as a template to produce RNA in a process called _____. (17.2)

24. Messenger RNA is used as a template to produce proteins in a process called _____. (17.2)

25. The first step of DNA replication is the _____ of the DNA helices. (17.2)

26. During DNA replication, new nucleotides are positioned and linked together by _____ _____ enzymes. (17.2)

27. DNA replication occurs at multiple _____ _____ along the length of mammalian chromosomes. (17.2)

28. Permanent changes that occur in a molecule of DNA are called _____. (17.2)

29. Mutations are required for _____ to occur. (17.2)

30. Errors in DNA replication are often repaired by DNA _____ _____. (17.2)

31. In the process of transcription, a base sequence called a _____ indicates where a gene begins. (17.2)

32. The enzyme RNA _____ positions nucleotides during transcription. (17.2)

33. A sequence of 3 bases that specify a particular amino acid is called a _____. (17.2)

34. The anticodon of a _____ molecule is complementary to a codon of a _____ molecule. (17.2)

35. Ribosomes are composed of _____ and _____. (17.2)

## Short Answer

36. Why does DNA replication in mammalian cells occur at multiple points along a chromosome at the same time?

37. List three things that can cause a mutation to occur in DNA.

38. How does a primary transcript differ from mRNA?

39. List the three types of RNA and the function of each type.

## Matching

*For each process depicted in Figure 17.2, identify the process and where in the cell it occurs.*

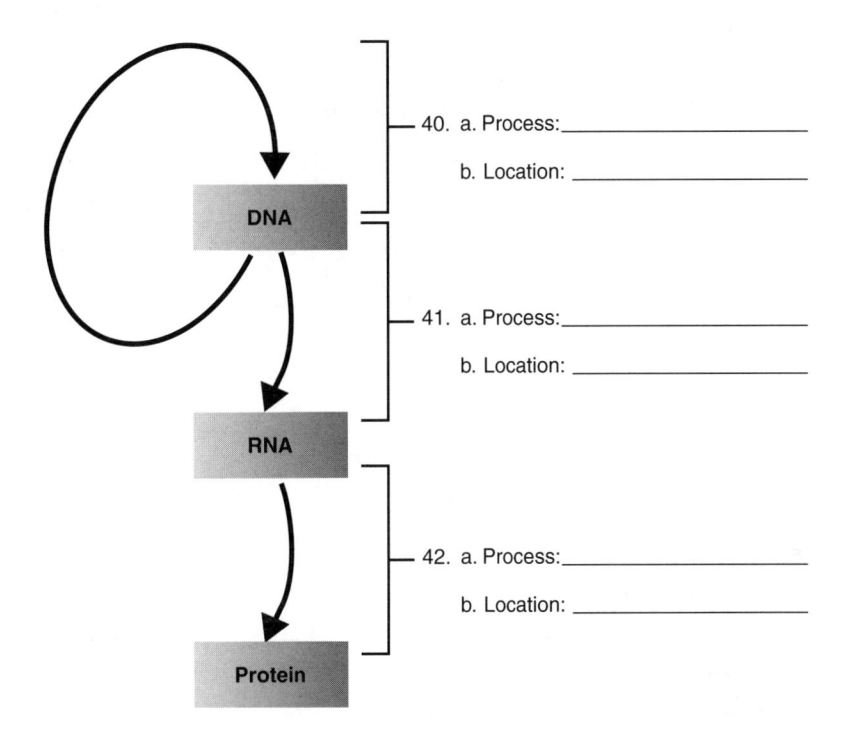

40. a. Process:_____
    b. Location:_____

41. a. Process:_____
    b. Location:_____

42. a. Process:_____
    b. Location:_____

**Figure 17.2**

## Completion

Use the base sequence of the DNA gene below to:

a. write the base sequence of the complementary DNA strand
b. write the base sequence of an mRNA strand produced from this gene
c. write the amino acid sequence of the protein produced from this gene, assuming this base sequence contains no introns

**DNA gene**                           **ACTACCCATGCCAG**

43. Complementary DNA strand: _____
44. Messenger RNA: _____
45. Protein amino acid sequence: _____

## Labeling

*Use the terms and events of translation listed below to label Figure 17.3.*

**Terms**
a. tRNA
b. start codon
c. amino acids
d. completed protein
e. stop codon
f. ribosomal subunits
g. mRNA
h. anticodon

**Events**
i. bond forms between amino acids
j. ribosome moves along mRNA
k. tRNA is released
l. tRNA captures free amino acids
m. elongation
n. termination
o. initiation

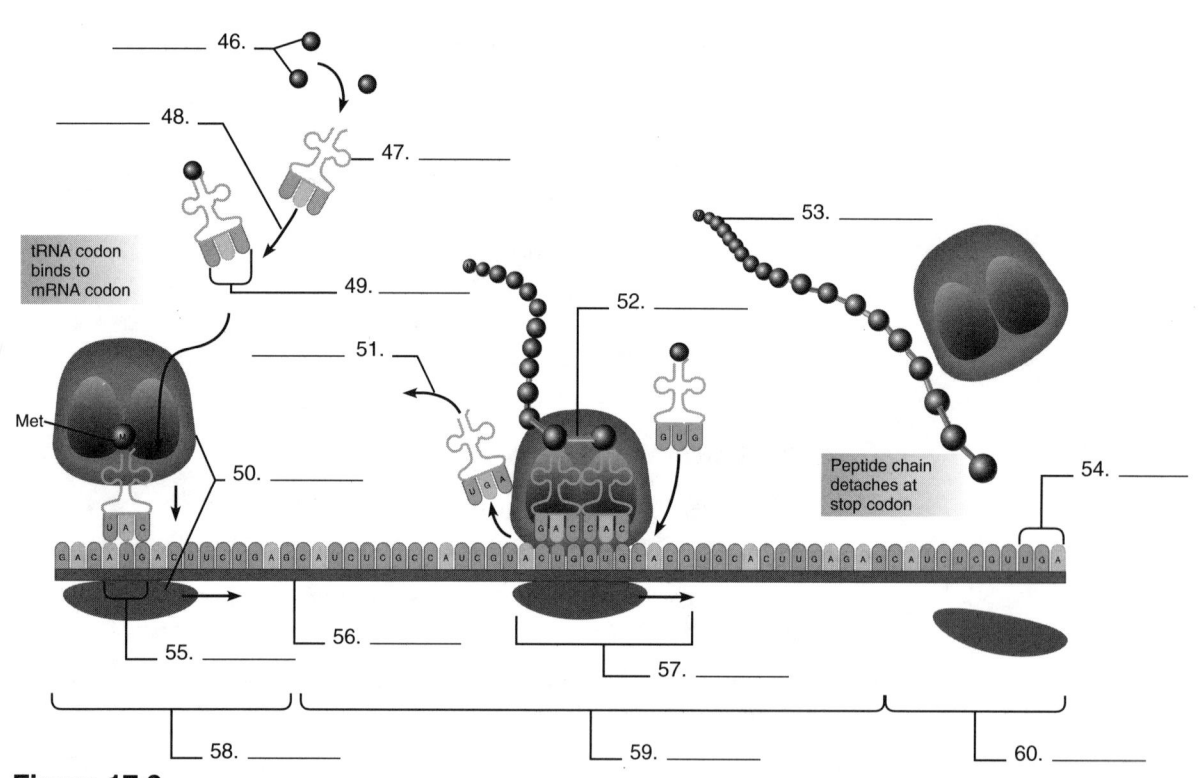

46. _____
47. _____
48. _____
49. _____
50. _____
51. _____
52. _____
53. _____
54. _____
55. _____
56. _____
57. _____
58. _____
59. _____
60. _____

**Figure 17.3**

17.3 Cell reproduction—one cell becomes two

17.4 How cell reproduction is regulated

17.5 Environmental factors influence cell differentiation

## Matching

____ 1. **mitosis**
____ 2. **cytokinesis**
____ 3. **prophase**
____ 4. **mitotic spindle**
____ 5. **metaphase**
____ 6. **anaphase**
____ 7. **telophase**
____ 8. **diploid cell**
____ 9. **homologous chromosomes**
____ 10. **haploid cell**
____ 11. **meiosis**
____ 12. **crossing-over**
____ 13. **differentiation**

a. a phase of mitosis when duplicated chromosomes separate and move toward opposite sides of the cell
b. the process whereby a cell becomes different from its parent or sister cells
c. a cell that contains both members of each homologous pair
d. division of the nucleus
e. cells that have only one member of each homologous pair
f. duplicated homologous chromosomes pair up and exchange sections of DNA during prophase of meiosis I
g. a cell division process that produces haploid cells
h. division of the cytoplasm
i. an arrangement of microtubules that facilitates equal division of the DNA
j. chromosomes that have copies of the same genes in the same locations
k. phase of mitosis when duplicated chromosomes first become visible
l. a phase of mitosis when the two sets of chromosomes arrive at opposite poles of the cell
m. a phase of mitosis when duplicated chromosomes align at the center of the cell

## Completion

*Place an "X" in the column under Mitosis, Meiosis I, or Meiosis II to indicate which cell division process includes the event described.*

| Event | Mitosis | Meiosis I | Meiosis II |
|---|---|---|---|
| 14. crossing-over occurs between homologous chromosomes | | | |
| 15. produces two haploid cells that still contain sister chromatids | | | |
| 16. daughter cells are diploid | | | |
| 17. in human cells 23 duplicated chromosomes will align on the equatorial plate | | | |
| 18. daughter cells are exactly like the parent cell | | | |
| 19. produces a secondary spermatocyte in a male | | | |
| 20. produces an egg in a female | | | |

## Labeling

*Use the terms and events of mitosis listed below to label Figure 17.4.*

**Terms**
a. cleavage furrow
b. mitotic spindle forming
c. daughter chromosomes
d. centrioles
e. mitotic spindle
f. centromere
g. duplicated chromosome

**Events**
h. anaphase
i. metaphase
j. telophase and cytokinesis
k. prophase
l. nuclear membrane forms
m. chromosomes uncoiling

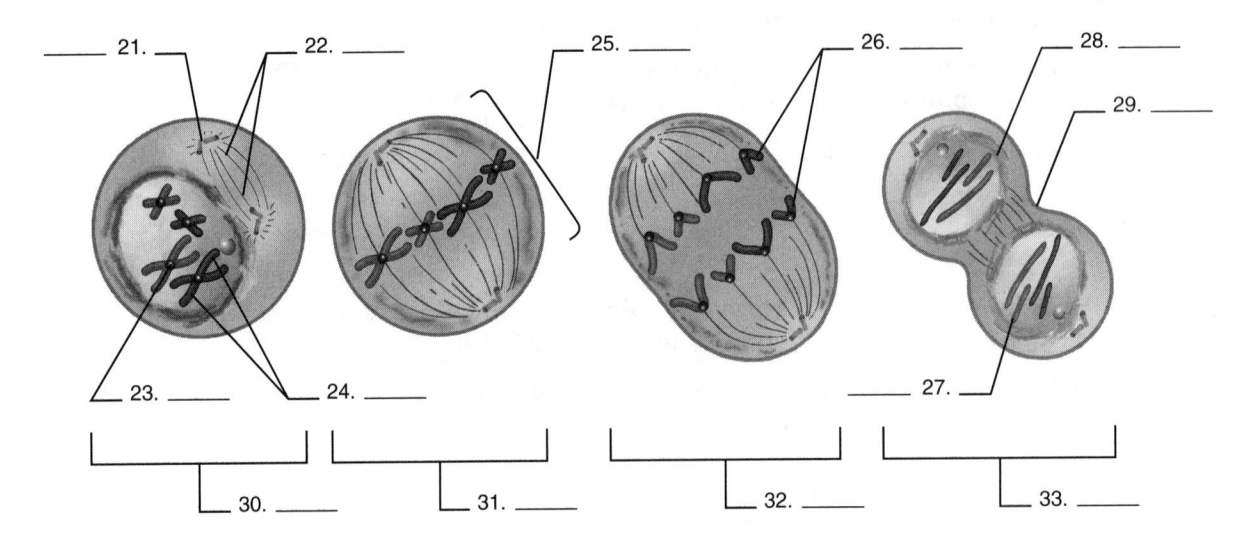

____ 21. ____
____ 22. ____
____ 25. ____
____ 26. ____
____ 28. ____
____ 29. ____
____ 23. ____
____ 24. ____
____ 27. ____
____ 30. ____
____ 31. ____
____ 32. ____
____ 33. ____

**Figure 17.4**

## Short Answer

*A human liver cell is diploid and has 46 chromosomes. Answer the following questions about a liver cell.*

34. How many chromosomes will be present in a liver cell in prophase?

35. How does the amount of DNA present in a prophase cell compare to the amount of DNA present in a cell in G1?

36. Does a cell in G2 contain duplicated chromosomes?

37. Is the DNA in a cell during G1 in the form of chromosomes or chromatin?

## Paragraph Completion

*Use the terms to complete the paragraph, then repeat this exercise with the terms covered.*

| hormone | checkpoints | identical | feedback control |
| eight | environmental | specialized | metaphase |
| $G_1$ | regulatory | cyclins | sixteen |
| $G_2$ | gene | parent | environment |
| nutrients | growth | inhibit | parent |

Cell division occurs at different rates in different cells. The rate at which a cell moves through the phases of the cell cycle is controlled by proteins called (38) _____. These proteins activate other proteins called (39) _____ proteins that trigger events during the cell cycle.

Cells also monitor events during the cell cycle at regulatory points called (40) _____. Events in one phase of the cycle must occur correctly before the cell will proceed to the next stage. The first important checkpoint occurs near the end of phase (41) _____. Additional checkpoints are located near the end of phase (42) _____ and in the (43) _____ stage of mitosis.

The cell cycle is also influenced by external conditions. Low availability of (44) _____ or the absence of a particular (45) _____ may interfere with progression of the cell cycle. Some cells require (46) _____ factors to be present. Tissue growth and organ size can be regulated when cells touch each other, causing the release of substances that (47) _____ cell division.

Cell differentiation occurs when cells become different than their (48) _____ cell. All the cells from one fertilized egg are genetically identical, and remain capable of forming a complete organism all the way through cell divisions that produce (49) _____ cells. Once about (50) _____ cells have been produced the cells are no longer (51) _____. This early differentiation is due to (52) _____ factors rather than (53) _____ mechanisms. Differentiation in later development is influenced by (54) _____ and the developmental history of the (55) _____ cell. Continued selective (56) _____ expression will create (57) _____ cells.

# Chapter Test

## Multiple Choice

1. Which phase of the cell cycle is incorrectly described?
    a. $G_1$—a time of slow cell growth
    b. S—sister chromatids are produced
    c. $G_2$—a cell prepares for cell division
    d. M—nuclear division occurs

2. Division of the cytoplasm occurs during:
   a. mitosis and cytokinesis.
   b. interphase.
   c. DNA replication.
   d. cytokinesis.

3. Histone proteins:
   a. contribute to the structure of a chromosome.
   b. become part of the mitotic spindle.
   c. disappear during prophase.
   d. direct chromosome movement during prophase.

4. A short segment of DNA that represents the code for a specific protein is a:
   a. codon.
   b. triplet.
   c. gene.
   d. chromosome.

5. When a protein has been produced from the information in a molecule of DNA, which of the following has occurred?
   a. replication
   b. transcription
   c. translation
   d. b and c
   e. all of the above

6. The enzyme involved in positioning and joining complementary bases during replication is:
   a. DNA replicase.
   b. DNA transcriptase.
   c. DNA polymerase.
   d. DNA nuclease.

7. A gene is expressed when:
   a. DNA is replicated.
   b. mutations are repaired.
   c. a protein is produced.
   d. sister chromatids are produced.

8. Which of the following is a way in which replication and transcription differ?
   a. Only replication requires complementary base pairing.
   b. Only transcription requires an enzyme.
   c. Transcription copies only one side of the DNA molecule.
   d. Replication occurs in the nucleus.

9. A final mRNA transcript will contain:
   a. introns and exons.
   b. introns only.
   c. exons only.
   d. a, b, or c, depending on the needs of the cell at the time.

10. _____ contains codons and _____ contains anticodons.
    a. tRNA, rRNA
    b. mRNA, tRNA
    c. mRNA, rRNA
    d. rRNA, tRNA

11. All proteins in the process of being produced will begin with the amino acid _____.
    a. Phe
    b. Ser
    c. Met
    d. every protein begins with a different amino acid, depending on the DNA message

12. In the process of translation, the instructions about the proper amino acid sequence of a new protein are carried by the:
    a. mRNA.
    b. tRNA.
    c. rRNA.
    d. ribosome.

13. During the phase of translation called elongation, the mRNA passes between _____, catalyzing the binding of _____ to each other.
    a. rRNA and tRNA, proteins
    b. two ribosomal subunits, amino acids
    c. two strands of DNA, nucleotides
    d. a ribosome and a tRNA, proteins

14. Which amino acid sequence is specified by the mRNA base sequence GAAUGCCCGCAC?
    a. glu-cys-pro-his
    b. leu-thr-gly-val
    c. met-pro-ala
    d. tyr-gly-arg

15. If the tubular elements of the cytoskeleton were not disassembled prior to mitosis, which of the following would be affected?
    a. formation of sister chromatids
    b. formation of the mitotic spindle
    c. disassembly of the nuclear membrane
    d. formation of a centromere

16. An original chromosome and its copy:
    a. float singly in the cytoplasm during mitosis.
    b. are joined to each other by a centromere.
    c. are attached down the entire length of the chromosome.
    d. form a pair of homologous chromosomes.
    e. become sister chromatids.
    f. b and e

17. During which phase of mitosis do individual chromosomes arrive at opposite poles of the cell?
    a. metaphase
    b. anaphase
    c. prophase
    d. telophase

18. Which of the following would not occur if the cleavage furrow was unable to form?
    a. DNA replication
    b. division of the centromere
    c. cytokinesis
    d. formation of the mitotic spindle

19. Meiosis in a male produces _____ cells, each of which is _____.
    a. two, haploid
    b. two, diploid
    c. four, haploid
    d. four, diploid
    e. one, haploid

20. Genetic recombination in meiosis is accomplished by a process called:
    a. mitosis.
    b. cleavage.
    c. homologous pairing.
    d. crossing-over.

21. A polar body is first produced:
    a. during meiosis I in a female.
    b. during meiosis II in a female.
    c. during meiosis I in a male.
    d. during meiosis II in a male.

22. A mutation in the DNA of an organism is passed on to offspring only when:
    a. it occurs during $G_2$ of the cell cycle.
    b. it occurs during the S phase of the cell cycle.
    c. it occurs in cells that give rise to an egg or a sperm.
    d. it affects both mitosis and cytokinesis.

23. Selective gene expression during development results in:
    a. genetic mutations.
    b. more advanced single celled organisms.
    c. the ability to clone an organism using cell separation.
    d. specialized cells.

24. Cells separated from each other after the 16-cell stage fail to produce an entire intact organism. This is because:
    a. They require influential chemicals from neighboring cells to develop.
    b. The trauma of separating cells at this stage prevents further development.
    c. DNA at the 16-cell stage is difficult to activate.
    d. The cells have already begun to differentiate.

25. Most human cells contain _____ chromosomes, while the cells of gametes have _____ chromosomes.
    a. 23, 23
    b. 46, 46
    c. 23, 46
    d. 46, 23

# Key Concept Review Questions

*Each of the Key Concepts listed at the beginning of this chapter has been rewritten as a question below. After successfully completing the study guide exercises and the Chapter Test, you should be able to answer each of these questions. Refer to the Key Concepts list at the beginning of this chapter to check your answers.*

1. What are the two primary phases of the cell cycle?
2. What three stages are included in interphase? What occurs during each stage?
3. What are the two divisions that make up the mitotic phase of the cell cycle?
4. How many chromosomes are found in human cells? What is a gene?
5. What is replication, and in what phase of the cell cycle does it occur?
6. Describe the structure of a replicated chromosome.
7. What two processes must occur for the information in DNA to be used to make a protein?
8. What is transcription and where in the cell does it occur?
9. What is translation and where in the cell does it occur?
10. What are the three types of RNA that participate in translation, and what is the function of each?
11. What is a mutation?
12. Are mutations always harmful? What process requires the presence of mutations?
13. What are the four phases of mitosis?
14. How many daughter cells are produced by mitosis? Are they identical to or different from the parent cell?
15. What are the two successive nuclear divisions of meiosis?
16. What is the number and ploidy (haploid or diploid) of the daughter cells produced by meiosis in a male and in a female?
17. How is the product of meiosis different in males and females?
18. What factors help to control the cell cycle?
19. Name two things that lead to the development of specialized cells.
20. What is differentiation?

# Answer Key

## Sections 17.1, 17.2

**1.** not labeled; **2.** d,g; **3.** m; **4.** b,f,g; **5.** n; **6.** c,e,f,g,i; **7.** p; **8.** f,g,k; **9.** l; **10.** a,h; **11.** o; **12.** f,g,j; **13.** interphase, mitotic phase; **14.** mitotic phase; **15.** cytokinesis; **16.** daughter; **17.** chromosomes; **18.** 46; **19.** histones; **20.** sister chromatids, centromere; **21.** gene; **22.** nucleus, pores; **23.** transcription; **24.** translation; **25.** uncoiling **26.** DNA polymerase; **27.** replication bubbles; **28.** mutations; **29.** evolution; **30.** repair enzymes; **31.** promoter; **32.** polymerase; **33.** codon; **34.** tRNA, mRNA;

**35.** rRNA, protein; **36.** to keep replication from taking too long; **37.** mistakes in DNA replication, chemicals, physical forces; **38.** the primary transcript contains introns and exons while the mRNA molecule contains only exons; **39.** mRNA carries the instructions on the correct sequence of amino acids in the protein, rRNA complexes with protein to become ribosomal subunits, tRNA transports amino acids from the cytoplasm to the ribosome during protein synthesis; **40.** a. replication, b. nucleus; **41.** a. transcription, b. nucleus; **42.** a. translation, b. cytoplasm; **43.** TGATGGGTACGGTC; **44.** UGAUGGGUACGGUC; **45.** met-gly-thr-val; **46.** c; **47.** a; **48.** l; **49.** h; **50.** f; **51.** k **52.** i; **53.** d; **54.** e; **55.** b; **56.** g; **57.** j; **58.** o; **59.** m; **60.** n

## Sections 17.3, 17.4, 17.5

**1.** d; **2.** h; **3.** k; **4.** i; **5.** m; **6.** a; **7.** l; **8.** c; **9.** j; **10.** e; **11.** g; **12.** f; **13.** b; **14.** Meiosis I; **15.** Meiosis I; **16.** Mitosis; **17.** Meiosis II; **18.** Mitosis; **19.** Meiosis I; **20.** Meiosis II; **21.** d; **22.** b; **23.** f; **24.** g; **25.** e; **26.** c; **27.** m; **28.** l; **29.** a; **30.** k; **31.** i; **32.** h; **33.** j; **34.** 46; **35.** a prophase cell has twice the amount of DNA because the chromosomes are replicated; each chromosome consists of two pieces of DNA; **36.** yes; **37.** chromatin; **38.** cyclins; **39.** regulatory; **40.** checkpoints; **41.** $G_1$; **42.** $G_2$; **43.** metaphase; **44.** nutrients; **45.** hormone; **46.** growth; **47.** inhibit; **48.** parent; **49.** eight; **50.** sixteen; **51.** identical; **52.** environmental; **53.** feedback control; **54.** environment; **55.** parent; **56.** gene; **57.** specialized

## Chapter Test

**1.** a; **2.** d; **3.** a; **4.** c; **5.** d; **6.** c; **7.** c; **8.** c; **9.** c; **10.** b; **11.** c; **12.** a; **13.** b; **14.** c; **15.** b; **16.** f; **17.** d; **18.** c; **19.** c; **20.** d; **21.** a; **22.** c; **23.** d; **24.** d; **25.** d

# 18

# Cancer: Uncontrolled Cell Division and Differentiation

## Chapter Summary and Key Concepts

*After reading and studying this chapter you should know the following:*

**Sections 18.1, 18.2, 18.3**

1. Cancer is a disease of cell division and differentiation.
2. Normal cells have carefully regulated cell division, and they generally stay in one location in the body.
3. Cancer cells exhibit unregulated cell division, and may move to distant locations in the body.
4. A tumor is a mass of cells that has lost control over cell division. Tumors may be benign or cancerous.
5. Benign tumors remain in one place as a well-defined mass of cells.
6. A tumor is considered a cancer when the cells lose their organization, structure, and regulatory control.
7. Cancerous tumors that remain in one place are in situ cancers.
8. Malignant tumors are formed by cancer cells that invade normal tissues and metastasize to distant sites.
9. In order for cancer to develop, cells must divide uncontrollably and break away from surrounding cells.
10. Proto-oncogenes are regulatory genes that promote normal cell growth.
11. Tumor suppressor genes are normal regulatory genes that suppress cell growth.
12. Oncogenes are genes that cause cancer and develop from mutated proto-oncogenes.
13. Mutated or damaged proto-oncogenes and tumor suppressor genes contribute to the development of cancer by causing or allowing abnormal cell division.
14. Substances that cause cancer are called carcinogens, and they include viruses and bacteria, environmental chemicals, tobacco, radiation, dietary factors, and alcohol.

15. Susceptibility to cancer can be inherited if parents pass oncogenes to their offspring.

16. Suppression of the immune system may allow some cancers to develop more easily.

**Sections 18.4, 18.5**

17. Early detection is crucial in treating cancer successfully.

18. Traditional cancer treatments include surgery, radiation, chemotherapy, or a combination of two or more treatments.

**Sections 18.6, 18.7**

19. The 10 most common cancers are lymphomas, leukemia, and cancers of the skin, breast, prostate, lung, colon and rectum, urinary bladder, uterus, and kidney.

20. Most cancers can be prevented by making wise lifestyle and nutritional choices.

# Exercises

Complete the exercises for each section after you have read and studied the section. If you cannot answer some questions, or answer them incorrectly, return to the chapter and review this information. You may find it helpful to work on only one section at a time. When you have completed all sections, take the Chapter Test as an indicator of your mastery of this topic.

18.1  **Tumors can be benign or cancerous**

18.2  **Cancerous cells lose control over cell functions**

18.3  **How cancer develops**

## Matching

___ 1. **hyperplasia**
___ 2. **tumor**
___ 3. **dysplasia**
___ 4. **cancer**
___ 5. **in situ cancer**
___ 6. **metastasis**
___ 7. **malignant tumor**
___ 8. **proto-oncogene**
___ 9. **oncogene**
___ 10. **tumor suppressor gene**
___ 11. **mutator gene**
___ 12. **carcinogenesis**
___ 13. **carcinogen**
___ 14. **melanoma**

a. the movement of cancer cells away from the main tumor
b. a substance or physical factor that causes cancer
c. a mutated proto-oncogene that contributes to the development of cancer
d. the process of transforming a normal cell into a cancerous cell
e. a condition that results from an abnormally high frequency of cell division
f. a type of cancer involving the pigment-producing cells of the skin
g. a mass of cells that has lost normal regulatory mechanisms over cell division; can be benign or cancerous
h. a normal regulatory gene that represses cell growth, division, differentiation, or adhesion
i. an unhealthy or abnormal change in cell structure
j. a general term for a tumor that contains some cells that have lost organization, structure, and regulatory control
k. a normal regulatory gene that promotes cell growth, differentiation, division, or adhesion
l. a cancerous tumor that remains in one place
m. cancers that invade normal tissue and metastasize to different sites
n. involved in DNA repair mechanisms during DNA replication.

## Fill-in-the-Blank

*Referenced sections are in parentheses.*

15. Hyperplasia often indicates that a cell has lost control over _____ _____. (18.1)

16. Tumors that stay in one place and are well-defined are usually _____. (18.1)

17. _____ is often a sign that tumors are precancerous. (18.1)

18. One in _____ people will die of cancer. (18.2)

19. Genes that code for proteins required for normal cell growth and differentiation are called _____ genes. (18.3)

20. Genes that code for proteins that influence the expression of structural genes are called _____ genes. (18.3)

21. _____ factors are proteins encoded by regulatory genes, and secreted by cells, that influence nearby cells. (18.3)

22. The development of cancer requires the presence of _____ oncogenes. (18.3)

23. The primary role of the gene p53 is to _____ cell division in cells that appear to be cancerous. (18.3)

24. Mutations in a _____ gene may lead to reduced DNA repair ability during DNA replication. (18.3)

25. The most significant risk factor for cancer may be _____. (18.3)

26. The _____ virus may contribute to liver cancer. (18.3)

27. The _____ _____ virus increases a woman's risk for cervical cancer. (18.3)

28. The most lethal carcinogen in the U.S. is _____. (18.3)

29. Dietary factors are probably involved in up to _____ % of all cancers. (18.3)

30. _____ _____ are reactive fragments of molecules produced during cellular metabolism that can damage healthy molecules in the body. (18.3)

31. Vitamins _____, _____, and _____ appear to have antioxidant benefits. (18.3)

32. Suppression of the _____ system may allow certain cancers to develop more easily. (18.3)

Chapter 18 *Cancer: Uncontrolled Cell Division and Differentiation* 273

## Labeling

*Use the terms and descriptions to label Figure 18.1. Some structures will have two terms or descriptions that apply.*

a. dysplasia
b. malignant tumor
c. hyperplasia
d. metastases
e. genetically altered epithelial cell
f. in situ cancer
g. cells stay in one place
h. cells change form
i. cancer cells invade normal tissue
j. cells divide more rapidly than normal

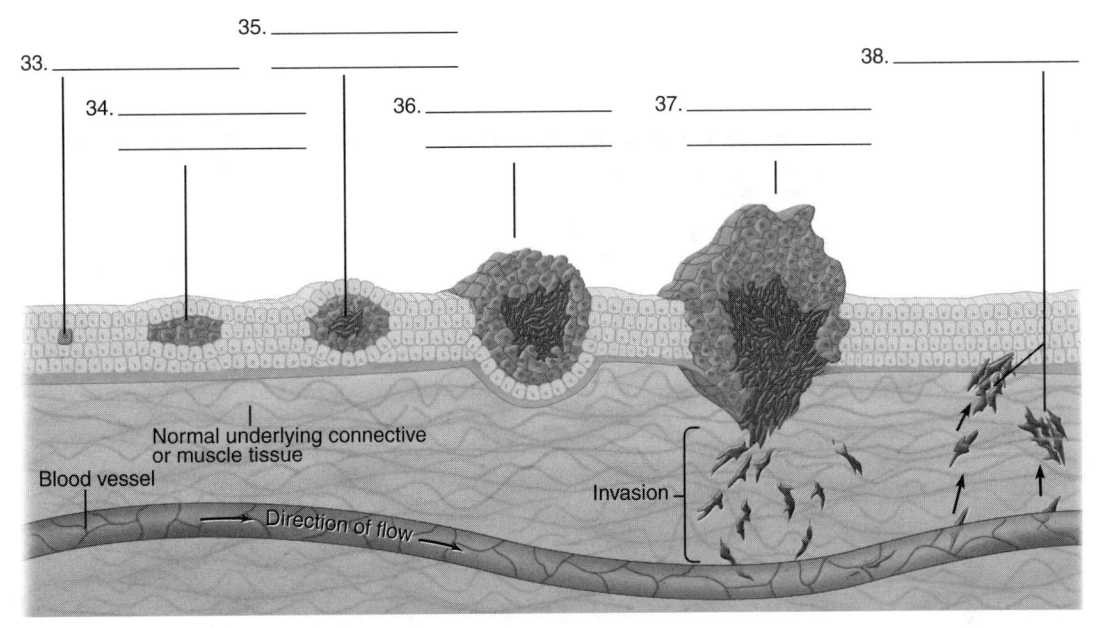

33. _____
34. _____
35. _____
36. _____
37. _____
38. _____

Normal underlying connective or muscle tissue
Blood vessel
Direction of flow
Invasion

**Figure 18.1**

## Short Answer

39. What do all types of cancer have in common?

40. List two characteristics of normal cells that are altered in cancer cells, and describe how they are altered.

41. What two things must happen simultaneously for cancer to develop?

42. What characteristics of viruses may contribute to their ability to cause cancer?

**18.4   Advances in diagnosis enable early detection**

**18.5   Cancer treatments**

## Completion

*Name each diagnostic method and cancer treatment described in the table.*

| Diagnostic Method or Treatment | Description |
|---|---|
| 1. | a diagnostic method in which radioactive substances are used to create three-dimensional images showing the metabolic activity of body structures |
| 2. | a diagnostic method which detects the presence of oncogenes |
| 3. | a diagnostic method in which body fluids are screened for the presence of telomerase |
| 4. | a cancer treatment involving the administration of cytotoxic chemicals to destroy cancer cells |
| 5. | a cancer treatment in which the cancer is targeted with lasers |
| 6. | a potential cancer treatment in which defective genes are repaired or replaced |
| 7. | a cancer treatment involving the use of anti-angiogenic drugs |

## Short Answer

8. Why are chemicals that inhibit cell division able to target primarily cancer cells while sparing most healthy cells?

9. How can the antigen of a cancer cell be used to develop immunotherapy treatments?

### 18.6 The 10 most common cancers

### 18.7 Most cancers can be prevented

## Short Answer

*Write the symptoms and risk factors for each cancer listed below. Answers for exercises in this section are not in the answer key; refer to the textbook to check your answers.*

1. Skin cancer:

2. Breast cancer:

3. Prostate cancer:

4. Lung cancer:

5. Colon and rectal cancer:

6. Lymphoma:

7. Urinary bladder cancer:

8. Uterine cancer:

9. Cervical cancer:

10. Kidney cancer:

11. Leukemia:

12. Using the recommendations in the textbook for cancer prevention, list at least six things you can do to reduce your risk of cancer:

    1. _____
    2. _____
    3. _____
    4. _____
    5. _____
    6. _____

---

# Chapter Test

## Multiple Choice

1. Cell division in healthy cells:
    a. is controlled by regulatory genes.
    b. is inhibited by signals from nearby cells.
    c. allows cells to travel through the body.
    d. a and b
    e. a, b, and c

2. Loss of regulatory control over cell division causes a cell to:
    a. grow unusually large.
    b. secrete inhibiting factors.
    c. divide more frequently than normal.
    d. all of the above.

3. Benign tumors threaten health only if they:
    a. secrete destructive enzymes.
    b. grow large enough to crowd normal cells.
    c. become cancerous.
    d. metastasize.

4. A tumor is defined as cancer when:
    a. it metastasizes.
    b. the cells have lost regulatory control over cell division.
    c. the cells have undergone structural changes.
    d. some of the cells have lost regulatory control over cell division and undergone structural changes.

5. Select the sequence of developments that occurs in forming a malignant tumor, beginning with a single cell.
   a. single mutated cell, dysplasia, hyperplasia, malignant tumor, in situ cancer
   b. single mutated cell, hyperplasia, dysplasia, in situ cancer, malignant tumor
   c. single mutated cell, in situ cancer, dysplasia, hyperplasia, malignant tumor
   d. single mutated cell, dysplasia, in situ cancer, hyperplasia, malignant tumor

6. Which of the following is the same in an in situ tumor and a malignant tumor?
   a. cell structure
   b. cell function
   c. frequency of cell division
   d. tumor organization

7. Which of the following does not occur as a cell progresses toward becoming cancerous?
   a. The nucleus may enlarge.
   b. There may be less cytoplasm.
   c. Cell function is impaired.
   d. The cell retains its normal structures.

8. Cancer ranks _____ as a cause of death.
   a. first
   b. second
   c. third
   d. fourth

9. Tumor suppresor genes are:
   a. normal regulatory genes that promote cell division.
   b. precancerous genes that have mutated at least once.
   c. normal regulatory genes that repress cell growth.
   d. regulatory genes that produce proteins that inhibit oncogenes.

10. Which of the following is *not* true of the gene p53?
    a. p53 encodes a gene that prevents damaged cells from dividing
    b. p53 is a tumor suppressor gene
    c. individuals who carry the p53 gene have an increased risk of cancer
    d. a mutated p53 gene has been found in some cervical cancers

11. A cell is dividing under circumstances when cell division would normally be inhibited. Which of the following is a regulatory gene that has probably been inactivated in the cell?
    a. a proto-oncogene
    b. an oncogene
    c. a tumor suppressor gene
    d. a growth factor

12. A cell has lost its ability to repair DNA errors that occur during DNA replication. What gene has probably been mutated?
    a. mutator gene
    b. oncogene
    c. proto-oncogene
    d. tumor suppressor gene

13. One mechanism by which viral infections may increase the risk of cancer is:
    a. viral infections increase the activity of tumor suppressor genes.
    b. the viral DNA sequence alters the function of a proto-oncogene.
    c. viral proteins interfere with maintenance of the plasma membrane.
    d. the virus interferes with the cell's ability to display "self" markers on the cell surface.

14. Sources of radiation that can increase cancer risk include:
    a. the sun.
    b. medical and dental x-rays.
    c. radon gas.
    d. all of the above.

15. Diet and tobacco together account for approximately _____ of all cancers.
    a. 10%
    b. 40%
    c. 60%
    d. 90%

16. Inheriting a mutated proto-oncogene from one or both of your parents means that:
    a. you will get cancer at a younger age than normal.
    b. you will get cancer, although it may be at an older age.
    c. you are at an increased risk for cancer.
    d. your risk for cancer is no different from anyone else's.

17. Which of the following statements about sunlight and skin cancer is true?
    a. Sunlamps and tanning booths are not as dangerous as natural sunlight.
    b. Sun exposure is only dangerous when it results in a sunburn.
    c. Any cumulative exposure to the sun increases the risk of cancer.
    d. Sunburns experienced during childhood do not increase the risk of cancer.

18. The presence of free radicals may increase the risk of cancer because:
    a. free radicals reduce a cell's ability to fight viral infections.
    b. free radicals increase sun damage to cells.
    c. free radicals can damage a cell's DNA.
    d. free radicals interfere with normal DNA repair mechanisms.

19. Antioxidant activity:
    a. inhibits the action of free radicals.
    b. is found in vitamin K.
    c. can reverse malignant changes in a cell.
    d. can convert a malignant tumor into a benign tumor.

20. In order for cancer to develop, what two things must happen?
    a. both a tumor suppressor gene and an oncogene must be mutated
    b. cells must divide uncontrollably, and they must experience physical changes that allow them to break away from surrounding cells
    c. dietary and environmental factors must be present that act as carcinogens
    d. a benign tumor must form, followed by hyperplasia

21. An enzyme that occurs rarely in normal cells but frequently in cancer cells is:
    a. prolactase.
    b. mutase.
    c. DNAse.
    d. telomerase.

22. The side effects of chemotherapy include nausea. This is because:
    a. the toxins in the drugs are difficult for the body to metabolize.
    b. rapidly dividing healthy cells, such as occur in the lining of the digestive tract, are also destroyed.
    c. chemotherapy interferes with the normal absorption of nutrients from the digestive tract.
    d. as the chemicals are metabolized, the by-products increase acid production in the stomach.

23. The growth of new blood vessels to support rapidly growing tumors is called:
    a. angiogenesis.
    b. cardiovascular supplementation.
    c. vascularization.
    d. capillary regeneration.

24. Which of the following skin cancers is incorrectly matched with its characteristic?
    a. basal cell cancer—metastasizes quickly
    b. squamous cell cancer—arises from epithelial cells
    c. melanoma—the least common form of skin cancer
    d. melanoma—the deadliest form of skin cancer

25. Cancer of the blood-forming organs:
    a. is called lymphoma.
    b. occurs more often in children than adults.
    c. is usually treated by radiation.
    d. may be caused by a virus.

# Key Concept Review Questions

*Each of the Key Concepts listed at the beginning of this chapter has been rewritten as a question below. After successfully completing the study guide exercises and the Chapter Test, you should be able to answer each of these questions. Refer to the Key Concepts list at the beginning of this chapter to check your answers.*

1. Cancer is a disease involving what two cellular processes?
2. What are the two primary characteristics of normal cells that are altered in cancer cells?
3. What are the two defining characteristics of cancer cells?
4. What characterizes the cell division of cells in a tumor? Are tumors always cancerous?
5. Describe a benign tumor.
6. What three things must tumor cells lose to be considered cancerous?
7. What is an in situ cancer tumor?
8. What is a malignant tumor?
9. What two things must happen in order for cancer to develop?
10. What are proto-oncogenes?
11. What are tumor suppressor genes?
12. What are oncogenes? How do oncogenes develop?

13. How do mutated proto-oncogenes or tumor suppressor genes contribute to the development of cancer?

14. What is a carcinogen? Name some common carcinogens.

15. How can susceptibility to cancer be inherited?

16. When does the immune system contribute to the development of cancer?

17. What is crucial in treating cancer successfully?

18. What are four traditional cancer treatments?

19. What are the 10 most common cancers?

20. Is cancer preventable?

# Answer Key

## Sections 18.1, 18.2, 18.3

**1.**e; **2.**g; **3.**i; **4.**j; **5.**l; **6.**a; **7.**m; **8.**k; **9.**c; **10.**h; **11.**n; **12.**d; **13.**b; **14.**f; **15.** cell division; **16.** benign; **17.** Dysplasia; **18.** four; **19.** structural; **20.** regulatory; **21.** Growth; **22.** multiple; **23.** inhibit; **24.** mutator **25.** age; **26.** hepatitis; **27.** human papilloma; **28.** tobacco; **29.** 30; **30.** Free radicals; **31.** A,C,E; **32.** immune **33.**e; **34.**c,j; **35.**a,h; **36.**f,g; **37.**b,i; **38.**d; **39.** All forms of cancer are diseases of cell division and differentiation **40.** Normal cells regulate cell division processes and stay in one location in the body; cancer cells lose control over cell division and may migrate to different areas of the body; **41.** For cancer to develop, cells must divide without control and gain the ability to break away from surrounding cells; **42.** Viruses insert their DNA into host cell DNA and may disrupt normal host cell gene function.

## Sections 18.4, 18.5

**1.** PET scan; **2.** genetic testing; **3.** enzyme testing; **4.** chemotherapy; **5.** photodynamic therapy; **6.** gene therapy; **7.** "starving" cancer; **8.** cancer cells divide rapidly, while most normal body cells do not undergo continuous cell division; **9.** the cancer antigen is used to produce antibodies that will target the cancer cell for destruction by the immune system

## Chapter Test

**1.**d; **2.**c; **3.**b; **4.**d; **5.**b; **6.**c; **7.**d; **8.**b; **9.**c; **10.**c; **11.**c; **12.**a; **13.**b; **14.**d; **15.**c; **16.**c; **17.**c; **18.**c; **19.**a; **20.**c; **21.**d; **22.**b; **23.**a; **24.**a; **25.**d

# 19

# Genetics and Inheritance

## Chapter Summary and Key Concepts

*After reading and studying this chapter you should know the following:*

**Sections 19.1, 19.2**

1. Genetics is the study of genes and their inheritance.
2. Human cells contain 22 pairs of chromosomes called autosomes and one pair of sex chromosomes.
3. Alleles are structural and functional variations of homologous genes.
4. An individual is homozygous when he or she possesses two identical alleles, and heterozygous when he or she possesses two different alleles.
5. An individual's complete set of genes and alleles is the genotype, and his or her observable physical and functional traits is the phenotype.
6. A Punnett square is a method for determining patterns of inheritance.
7. Gregor Mendel proposed the basic rules of inheritance in the 1850s.
8. The law of segregation states that gametes receive only one allele of each gene pair.
9. The law of independent assortment states that genes for different traits are separated from each other during meiosis.
10. In complete dominance, a dominant allele masks the effect of a recessive allele.
11. In incomplete dominance, the heterozygous genotype produces a phenotype that is intermediate between the two homozygous phenotypes.
12. In codominance, both alleles present will be equally expressed.
13. In polygenic inheritance, a phenotypic trait depends on the action of multiple genes.

14. Many phenotypes are determined by genotype and environmental effects.

15. Linked genes are genes located on the same chromosome and may be inherited together.

**Sections 19.3, 19.4, 19.5, 19.6**

16. A non-sex gene located on a sex chromosome shows a pattern of sex-linked inheritance.

17. A trait that is not carried on a sex chromosome but is influenced by the expression of a sex chromosome is a sex-influenced trait.

18. Chromosome structure or number may be altered by nondisjunction, deletions, and translocations.

19. Nondisjunction occurs when homologous chromosomes or sister chromatids fail to separate during cell division, a deletion occurs when a piece of a chromosome breaks off and is lost, and a translocation occurs when a piece of a chromosome breaks off and reattaches at another site.

20. Inherited genetic disorders usually involve recessive alleles.

# Exercises

*Complete the exercises for each section after you have read and studied the section. If you cannot answer some questions, or answer them incorrectly, return to the chapter and review this information. You may find it helpful to work on only one section at a time. When you have completed all sections, take the Chapter Test as an indicator of your mastery of this topic.*

**19.1   Your genotype is the genetic basis of your phenotype**

**19.2   Genetic inheritance follows certain patterns**

## Matching

\_\_\_\_  1. **genetics**         a. a heritable change in the DNA

\_\_\_\_  2. **inheritance**      b. a type of inheritance in which the presence of one dominant allele is sufficient to cause expression of the dominant phenotype

\_\_\_\_  3. **autosome**         c. genes that are on the same chromosome

\_\_\_\_  4. **sex chromosomes**  d. an allele that masks the expression of the partner allele on the homologous chromosome

\_\_\_\_  5. **allele**           e. the study of genes

\_\_\_\_  6. **homozygous**       f. the observable physical and functional traits of an individual

_____ 7. **heterozygous**     g. an allele that must be present on both homologous chromosomes to be expressed

_____ 8. **mutation**     h. a method used to predict patterns of inheritance

_____ 9. **genotype**     i. a type of inheritance in which the heterozygous genotype results in a phenotype intermediate between the two homozygous phenotypes

_____ 10. **phenotype**     j. a type of inheritance in which the expression of a phenotypic trait depends on many genes

_____ 11. **Punnett square**     k. chromosomes that determine gender

_____ 12. **dominant allele**     l. something received from an ancestor

_____ 13. **recessive allele**     m. a condition in which both alleles for a particular gene are identical

_____ 14. **complete dominance**     n. a chromosome that does not determine gender

_____ 15. **incomplete dominance**     o. the complete set of genes and alleles possessed by an individual

_____ 16. **codominance**     p. a structural and functional variation of a homologous gene

_____ 17. **polygenic inheritance**     q. a condition in which the alleles for a particular gene are different

_____ 18. **linked genes**     r. a type of inheritance in which both alleles present are fully and equally expressed

## Fill-in-the-Blank

*Referenced sections are in parentheses.*

19. Humans carry _____ copy(ies) of each inherited gene. (19.1)

20. Different alleles that exist in a genome probably arose as a result of _____. (19.1)

21. The complete set of genes on the chromosomes of a particular organism is called the _____. (19.1)

22. The basic rules of inheritance were first proposed by an Austrian monk named _____ _____. (19.2)

23. The Law of Segregation states that two alleles will be separated during meiosis so that each gamete receives _____ copy(ies). (19.2)

24. The law of _____ _____ states that genes for different traits are separated from each other during the formation of gametes. (19.2)

25. Phenotype is affected by both the _____ and the _____. (19.2)

26. _____ alleles are represented by an upper case letter. (19.2)

27. Genes that are close together on a chromosome and are inherited together are called _____ _____. (19.2)

28. The "factors" of heredity proposed by Mendel are now known to be _____. (19.2)

## Labeling

*Write in the genotypes of the male and female parent in Figure 19.1, and then complete the Punnett square.*

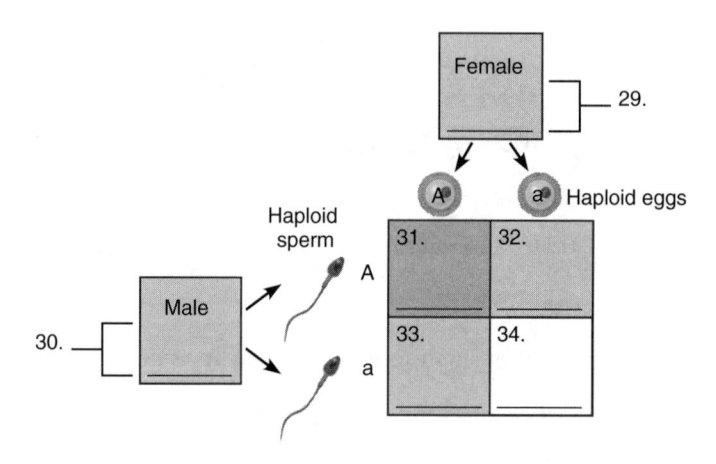

**Figure 19.1**

## Short Answer

35. Albinism, a trait in which individuals have no pigment, is inherited as a recessive trait (a), while normal pigment is carried by the dominant allele (A). Only individuals who are homozygous will be albino.
    a. What is the genotype of an albino individual?
    b. What is the phenotype of an individual whose genotype is Aa?
    c. What is the genotype of an individual who is heterozygous?

d. What is the phenotype of an individual who is homozygous dominant?
e. Use a Punnett square to determine the possible genotypes of all children born to a heterozygous father and an albino mother. What percentage of the children will be expected to be albino?

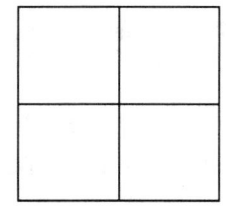

36. In the inheritance of hair type, homozygous dominant (WW) individuals have curly hair, homozygous recessive (ww) individuals have straight hair, and heterozygous (Ww) individuals have wavy hair.
    a. What type of inheritance is occurring?
    b. Use a Punnett square to determine the genotypes of the children born to a curly-haired father and a straight-haired mother.

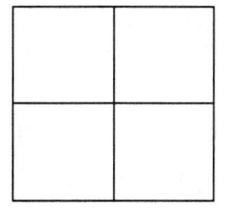

37. The shape of the hemoglobin molecule in a red blood cell is a genetic trait. The normal hemoglobin gene ($Hb^A$) codes for a normal shaped red blood cell, while the sickle cell gene ($Hb^S$) codes for an abnormally shaped hemoglobin molecule that becomes sickle-shaped during physiological stress. Sickle cell is inherited in a codominant fashion, although the phenotype appears more like an incompletely dominant trait. Individuals who carry two normal genes have normal red blood cells, individuals who carry one of each have Sickle Cell trait with some hemoglobin molecules of each type, and individuals who carry two sickle cell genes have Sickle Cell Anemia and no normal hemoglobin molecules.
    a. What is the phenotype of an individual who is heterozygous?
    b. What is the genotype of an individual with Sickle Cell Anemia?
    c. Use a Punnett square to determine the possible genotypes of all children born to a father with Sickle Cell trait, and a mother who has Sickle Cell Anemia. What percentage of the children can be expected to have only normal hemoglobin molecules? Why is this a codominant trait, and why does the phenotype appear more like a trait that is incompletely dominant?

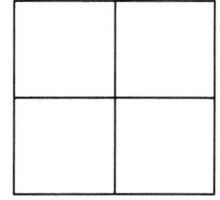

19.3 Sex-linked inheritance: X and Y chromosomes carry different genes

19.4 Chromosomes may be altered in number or structure

19.5 Many inherited genetic disorders involve recessive alleles

19.6 Genes code for proteins, not for specific behaviors

## Fill-in-the-Blank

*Referenced sections are in parentheses.*

1. A composite display of all of the chromosomes of an organism is called a _____. (19.3)

2. Human cells contain 22 pairs of chromosomes called _____ and one pair of _____ chromosomes. (19.3)

3. In humans, maleness is determined by the presence of a _____ chromosome. (19.3)

4. An inheritance pattern that depends on genes located on sex chromosomes is called _____-_____ inheritance. (19.3)

5. A trait that is not inherited on a sex chromosome but is influenced by the expression of a sex chromosome is called a _____-_____ trait. (19.3)

6. _____ occurs when homologous chromosomes or sister chromatids fail to separate during meiosis. (19.4)

7. A _____ occurs when a piece of a chromosome breaks off and is lost. (19.4)

8. A _____ occurs when a piece of a chromosome breaks off and reattaches at another site. (19.4)

9. The goal of the _____ _____ project is to identify the location and function of all genes in the human genome. (19.6)

## Short Answer

10. Why do more males than females have X-linked disorders?

11. Why can a man inherit an X-linked trait only from his mother?

12. Why will all the daughters of a man with an X-linked trait be carriers for the disorder?

13. Hemophilia is inherited as an X-linked recessive trait. Use $X^H$ to represent the allele for normal blood, and $X^h$ to represent the allele for hemophilia.
    a. What is the genotype of a man who has hemophilia?
    b. What is the genotype of a woman who has hemophilia?
    c. What is the genotype of a man who has normal blood?
    d. What is the genotype of a woman who has normal blood?
    e. What is the genotype of a woman who has normal blood and is a carrier of hemophilia?
    f. Use a Punnett square to determine the possible genotypes of children born to a man with hemophilia and a woman who has normal blood but is a carrier of the trait. What percentage of their sons can be expected to have normal blood? What percentage of their daughters can be expected to have hemophilia?

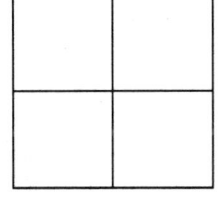

14. Why are sex-influenced traits, such as pattern baldness, expressed differently in men and women who both have the heterozygous genotype?

15. Why might the ability of genes to encode proteins influence behavior and personality?

## Completion

*Identify each genetic disorder described below.*

| Genetic Disorder | Description |
|---|---|
| 16. | Individuals with this disorder have three X chromosomes, are normal females, and often have mild retardation. |
| 17. | This disorder is inherited as an autosomal recessive and develops because affected individuals are unable to metabolize phenylalanine. |
| 18. | This disorder is caused by a deletion in chromosome 5, and affected infants are usually mentally and physically retarded, with a cat-like cry. |
| 19. | Affected men have the genotype XXY, are tall and sterile, and may have mild mental impairment. |
| 20. | This disorder is most often caused by an extra copy of chromosome 21. |
| 21. | This disorder is caused by a dominant-lethal allele and results in progressive nerve degeneration. |

## Labeling

*Draw the chromosome in the cells below indicating the results of nondisjunction at different stages of meiosis. (Refer to Figure 19.14 of the textbook to check your answers.)*

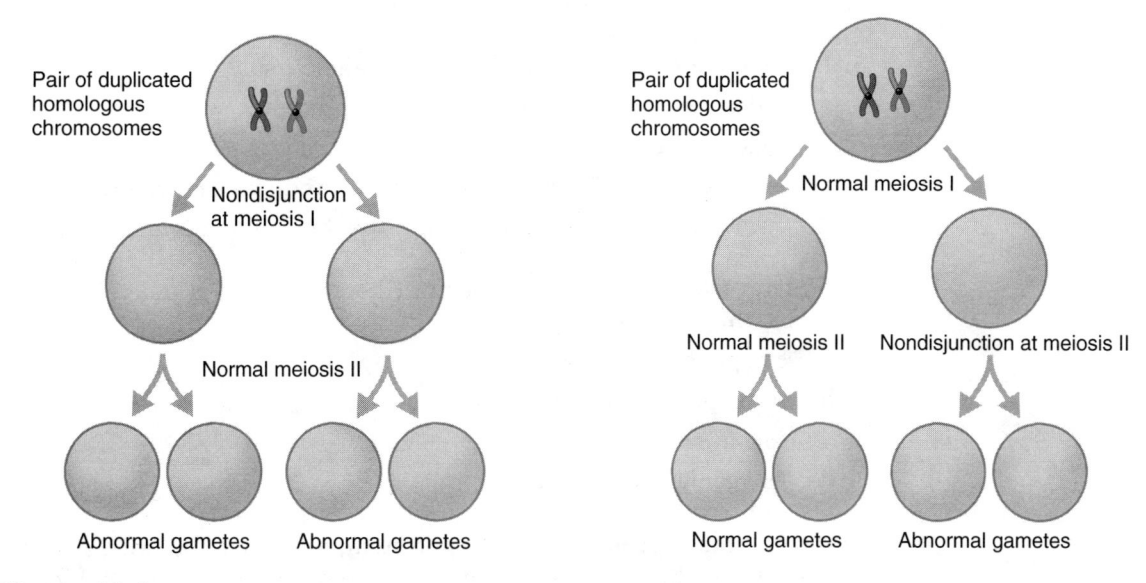

**Figure 19.2**

---

# Chapter Test

## Multiple Choice

1. Human cells contain 22 pairs of _____.
   a. chromosomes
   b. alleles
   c. homologues
   d. autosomes

2. An individual whose genotype is Ww is referred to as being:
   a. homozygous.
   b. heterozygous.
   c. genotypic.
   d. dominant.

3. The law of segregation states that:
   a. in the formation of gametes, alleles separate from each other so that each gamete receives only one allele.
   b. male and female gametes segregate independently of each other.
   c. the genes on one chromosome move independently of the genes on another chromosome.
   d. only alleles that are not linked will be segregated during the formation of gametes.

4. A recessive phenotype will be expressed when:
   a. at least one recessive allele is present.
   b. at least one dominant allele is present.
   c. both alleles present are recessive.
   d. both alleles present are dominant.

5. In complete dominance, the recessive allele usually:
   a. codes for only a small amount of a protein.
   b. codes for a fully functional protein.
   c. codes for a nonfunctional protein or produces no protein.
   d. codes for a protein that is repressed by the dominant allele.

6. When a white horse and a chestnut horse produce a palomino horse, this is an example of:
   a. complete dominance.
   b. incomplete dominance.
   c. codominance.
   d. polygenic inheritance.

7. In the human ABO blood-typing system, an individual with A and B alleles will express both A and B phenotypes on their red blood cells. This is an example of:
   a. complete dominance.
   b. incomplete dominance.
   c. codominance.
   d. polygenic inheritance.

8. In humans, the allele for a widow's peak (W) is dominant. Which of the following represent the possible phenotypes in the offspring of two parents, both with a straight hairline?
   a. All offspring will have a widow's peak.
   b. All offspring will have a straight hairline.
   c. Offspring will show both phenotypes.
   d. Most of the offspring will have a straight hairline, but it is possible to have a widow's peak.

9. When continuous variation occurs in a phenotypic trait, the trait is likely due to:
   a. complete dominance.
   b. incomplete dominance.
   c. codominance.
   d. polygenic inheritance.

10. The frequency with which genes on a particular chromosome are inherited together depends on:
    a. whether they are dominant or recessive.
    b. the occurrence of nondisjunction during meiosis.
    c. how close together they are on the chromosome.
    d. how often crossing over occurs.
    e. c and d

11. In humans, femaleness is determined by:
    a. the presence of at least one X chromosome.
    b. the presence of two X chromosomes.
    c. the absence of a Y chromosome.
    d. the presence of adequate levels of estrogen.

12. Testosterone is responsible for the development of the male reproductive system, and it is produced by the testis. The development of the testes depends on the presence of a Y chromosome. What might occur if the Y chromosome is present, but the testes are unable to produce testosterone?
    a. normal male development
    b. normal female development
    c. the testes would be present, but further male development would be arrested
    d. testes and ovaries would be present

13. Hemophilia is an X-linked recessive trait resulting in the inability to clot the blood. A man with hemophilia inherited the condition from:
    a. his mother.
    b. his father.
    c. both of his parents.
    d. all of the above could produce hemophilia in a man.

14. A woman with hemophilia inherited the condition from:
    a. her mother.
    b. her father.
    c. both of her parents.
    d. all of the above could produce hemophilia in a woman.

15. Baldness is a sex-influenced trait. Women must inherit both recessive alleles to be bald, while men need only inherit one recessive allele. This occurs because:
    a. estrogen masks the effect of the baldness allele.
    b. the presence of two X chromosomes in a woman masks the baldness allele.
    c. the baldness allele is linked to the Y chromosome.
    d. testosterone stimulates the expression of the baldness allele.

16. Nondisjunction in meiosis I results in failure of:
    a. homologous chromosomes to separate.
    b. sister chromatids to separate.
    c. gametes to separate.
    d. centromeres to separate.

17. An infant born with a deletion in chromosome 5 will have:
    a. Down syndrome.
    b. Turner syndrome.
    c. Edwards' syndrome.
    d. cri du chat syndrome.

18. An infant born with the genotype XYY will:
    a. be phenotypically female and fertile.
    b. be phenotypically male, short, and mildly retarded.
    c. be phenotypically male and usually normal.
    d. show partial male and female development.

19. Dominant alleles that cause disease are especially dangerous because:
    a. the severity of the disease is greater than those caused by recessive alleles.
    b. heterozygous individuals will develop the disease.
    c. individuals must inherit two dominant alleles to show the disease.
    d. dominant alleles are inherited more frequently than recessive alleles.

20. Which of the following is not a source of genetic variability during sexual reproduction?
    a. independent assortment
    b. crossing over between homologous chromosomes
    c. polygenic inheritance
    d. random fertilization of an egg by a sperm

21. Alterations in chromosomal structure or number that occur during mitosis or meiosis are seldom seen because:
    a. they do not affect the organism.
    b. the effects are so mild that they often go unrecognized.
    c. the effects are usually masked by more serious conditions.
    d. the altered genes are crucial to development and the organism dies.

22. A man carries a Y-linked trait. He will pass this trait to:
    a. all of his daughters.
    b. all of his sons.
    c. 50% of his daughters and 100% of his sons.
    d. 50% of his sons and none of his daughters.

23. When a translocation occurs:
    a. a piece of a chromosome breaks off and is lost.
    b. a damaged chromosome is passed from parent to child.
    c. an X-linked trait is passed to a son.
    d. a segment of a chromosome breaks off and reattaches at another location.

24. Which of the following is a condition caused by a dominant-lethal allele?
    a. Down syndrome
    b. Huntington disease
    c. Klinefelter syndrome
    d. Hemophilia

25. Specific genes code for specific:
    a. proteins.
    b. behaviors.
    c. traits.
    d. allele combinations.

# Key Concept Review Questions

*Each of the Key Concepts listed at the beginning of this chapter has been rewritten as a question. After successfully completing the study guide exercises and the Chapter Test, you should be able to answer each of these questions. Refer to the Key Concepts list at the beginning of this chapter to check your answers.*

1. What is genetics?
2. How many pairs of chromosomes, and of what type, do human cells contain?
3. What are alleles?
4. What makes an individual homozygous or heterozygous?
5. How would you define genotype and phenotype?
6. What is a Punnett square used for?
7. Who proposed the basic rules of inheritance, and in what year?
8. What principle is stated by the law of segregation?
9. What principle is stated by the law of independent assortment?
10. What occurs in complete dominance?
11. How is incomplete dominance different from complete dominance?
12. How are alleles expressed in codominance?
13. What determines phenotype in polygenic inheritance?

14. What, besides genotype, can influence the phenotypic expression of a trait?

15. What are linked genes?

16. What causes sex-linked inheritance?

17. What is a sex-influenced trait?

18. What are three ways in which chromosome number or structure can be altered?

19. What occurs in nondisjunction? What is a deletion? What is a translocation?

20. Do inherited disorders usually involve recessive or dominant alleles?

## Answer Key

### Sections 19.1, 19.2

**1.**e; **2.**l; **3.**n; **4.**k; **5.**p; **6.**m; **7.**q; **8.**a; **9.**o; **10.**f; **11.**h; **12.**d; **13.**g; **14.**b; **15.**i; **16.**r; **17.**j; **18.**c; **19.** two; **20.** mutation; **21.** genome; **22.** Gregor Mendel; **23.** one; **24.** independent assortment; **25.** genotype, environment; **26.** Dominant; **27.** linkage group; **28.** genes; **29.** Aa; **30.** Aa; **31.** AA; **32.** Aa; **33.** Aa; **34.** aa; **35.** a. aa, b. normal pigmentation, c. Aa, d. normal pigmentation, e. 50%;

|   | a | a |
|---|---|---|
| A | Aa | Aa |
| a | aa | aa |

**36.** a. incomplete dominance, b.

|   | W | W |
|---|---|---|
| w | Ww | Ww |
| w | Ww | Ww |

**37.** a. Hb$^A$Hb$^S$, b. Hb$^S$Hb$^S$, c. 0%, Sickle Cell is codominant because both traits are expressed when both alleles are present. It appears almost like an incompletely dominant trait because the heterozygous phenotype is intermediate between the fully dominant and fully recessive phenotypes,

|   | Hb$^S$ | Hb$^S$ |
|---|---|---|
| Hb$^A$ | Hb$^A$Hb$^S$ | Hb$^A$Hb$^S$ |
| Hb$^S$ | Hb$^S$Hb$^S$ | Hb$^S$Hb$^S$ |

### Sections 19.3, 19.4, 19.5, 19.6

**1.** karyotype; **2.** autosomes, sex; **3.** Y; **4.** sex-linked; **5.** sex-influenced; **6.** Nondisjunction; **7.** deletion; **8.** translocation; **9.** Human Genome; **10.** males inherit only one X chromosome; they cannot be protected by a normal allele on a second X chromosome; **11.** men receive a Y chromosome from their father; their mother is the only parent that donates an X chromosome; **12.** the father has only one X chromosome—the one with the trait—and he will pass it to all of his daughters; **13.** a. $X^hY$, b. $X^hX^h$, c. $X^HY$, d. $X^HX^H$ or $X^HX^h$, e. $X^HX^h$, f. 50%, 50%

|     | $X^H$     | $X^h$     |
|-----|-----------|-----------|
| $X^h$ | $X^HX^h$ | $X^hX^h$ |
| Y   | $X^HY$   | $X^hY$   |

**14.** The expression of pattern baldness, a sex-influenced trait, is enhanced by testosterone causing it to be expressed more strongly in men than in women; **15.** proteins play many roles in body chemistry that could influence behavior; **16.** Triple X syndrome; **17.** PKU; **18.** cri-du-chat syndrome; **19.** Klinefelter syndrome; **20.** Down syndrome; **21.** Huntington disease

## Chapter Test

**1.**d; **2.**b; **3.**a; **4.**c; **5.**c; **6.**b; **7.**c; **8.**b; **9.**d; **10.**e; **11.**c; **12.**c; **13.**a; **14.**c; **15.**d; **16.**a; **17.**d; **18.**c; **19.**b; **20.**c; **21.**d; **22.**b; **23.**d; **24.**b; **25.**a

# 20

# DNA Technology and Genetic Engineering

## Chapter Summary and Key Concepts

*After reading and studying this chapter you should know the following:*

### Sections 20.1, 20.2

1. Biotechnology is the technical application of biological knowledge for human purposes.

2. DNA sequencing allows scientists to determine the base sequence of a DNA molecule.

3. DNA sequencing relies on complementary base pairing to produce a new DNA strand, and requires the use of short single-stranded DNA segments, primers, modified DNA nucleotides, DNA polymerase, gel electrophoresis, and a fluoroscope.

4. Recombinant DNA is produced by combining DNA from two different sources. Recombinant DNA technology provides tools for manipulating DNA, and involves cutting, splicing, and copying DNA.

5. Recombinant bacteria are produced using DNA from a foreign organism, restriction enzymes, DNA ligases, and plasmids.

6. Restriction enzymes cut DNA by breaking the bonds between specific base pairs.

7. DNA ligases can bind DNA fragments from different organisms together after they have been cut by restriction enzymes.

8. Plasmids are extrachromosomal DNA molecules that occur naturally in bacteria.

9. Plasmids can accept a foreign piece of DNA and carry it back into the bacteria. Bacteria containing a foreign gene can manufacture a foreign protein.

10. Polymerase chain reaction (PCR) is a technique that amplifies a small fragment of DNA.

11. PCR uses separated strands of a small piece of DNA to serve as a template for the construction of new complementary strands. Cycles of heating and cooling allow the process to repeat until many identical DNA molecules are produced.

**Sections 20.3, 20.4**

12. Transgenic organisms carry foreign genes as a result of genetic engineering.

13. Transgenic bacteria created with engineered plasmids can produce foreign proteins, including human proteins.

14. Transgenic plants carry foreign genes and have been engineered to resist adverse weather and pests, carry vitamin precursors, and produce vaccines.

15. Transgenic plants have created some concerns about potential safety and environmental hazards associated with their use.

16. Creating transgenic animals is difficult, but research continues because of the great potential benefits.

17. Current techniques for creating transgenic animals include inserting DNA into a single isolated cell, which can give rise to a whole organism.

18. Gene therapy involves the insertion of human genes into human cells for medical purposes.

19. Gene therapy may utilize retroviruses as vectors—agents that can transport a gene into a human cell.

20. Gene therapy has been successfully used to treat Severe Combined Immune Deficiency, and research continues in the treatment of cystic fibrosis and cancer.

# Exercises

*Complete the exercises for each section after you have read and studied the section. If you cannot answer some questions, or answer them incorrectly, return to the chapter and review this information. You may find it helpful to work on only one section at a time. When you have completed all sections, take the Chapter Test as an indicator of your mastery of this topic.*

**20.1 DNA sequencing reveals structure of DNA**

**20.2 DNA can be modified in the laboratory**

## Matching

____ 1. **biotechnology**  a. enzymes that bind DNA fragments together

____ 2. **recombinant DNA technology**  b. enzymes that cut DNA into fragments by breaking the bonds between specific base pairs

____ 3. **genetic engineering**  c. short single-stranded pieces of DNA that bind to a single strand of DNA and serve as a starting point for the synthesis of a new strand

____ 4. **primers**  d. a process that can amplify a small sample of DNA

____ 5. **DNA polymerase**  e. a process that separates DNA fragments according to size

____ 6. **gel electrophoresis**  f. the technical application of biological knowledge for human purposes

____ 7. **restriction enzymes**  g. small, circular, self-replicating DNA molecules found in bacteria

____ 8. **palindrome**  h. an enzyme required for DNA synthesis

____ 9. **DNA ligases**  i. tools that allow cutting, splicing, and copying DNA

____ 10. **plasmids**  j. manipulation of the genetic makeup of cells or whole organisms

____ 11. **polymerase chain reaction**  k. a sequence of DNA bases often recognized and cut by restriction enzymes

## Completion

*Use the terms below to complete the paragraph, then repeat this exercise with the terms covered.*

| smaller | strands | test tube |
| fluorescent | labeled | fluoroscope |
| labeled | synthesis | base |
| polymerase | synthesis | lengths |
| primers | size | larger |
| complementary | original | gel electrophoresis |

The process of DNA sequencing allows biologists to determine the (12) _____ sequence of a strand of DNA. First, many copies of a short, single-stranded piece of DNA are placed in a (13) _____. Next, (14) _____ are added. Primers are short, single-stranded pieces of DNA that bind to one end of the (15) _____. A primer is needed to provide a starting point for the (16) _____ of a new complementary DNA strand. Also required are free, available DNA nucleotides, containing the bases A, T, C, and G. A mixture of normal nucleotides and (17) _____ nucleotides will be added to the tube. The labeled nucleotides are identifiable because they have a (18) _____ marker. Labeled nucleotides have also been altered so that DNA (19) _____ will stop as soon as they are added to the chain. Finally, the enzyme DNA (20) _____ will be added; this enzyme is required for DNA synthesis. When the process is completed, many DNA strands of different (21) _____ will have been produced, each ending with a (22) _____ nucleotide. A process called (23) _____ sorts the strands by (24) _____.

As the strands move through the gel column, the (25) _____ pieces move more quickly than the (26) _____ pieces. As each piece of DNA moves off the gel a (27) _____ detects the fluorescent markers. This information is used to sequence the bases of the original DNA strand; base pairing rules allow scientists to then determine the base sequence of the (28) _____ strand.

## Fill-in-the-Blank

*Referenced sections are in parentheses.*

29. Every organism produced by sexual reproduction contains _____ DNA. (20.2)

30. Restriction enzymes cut DNA between specific _____ _____. (20.2)

31. In a palindromic DNA sequence, if one strand of DNA has the base sequence CTGCAG, the base sequence of the complementary strand will be _____. (20.2)

32. A plasmid is an extrachromosomal, self-replicating DNA molecule found in _____. (20.2)

33. Plasmids contain genes needed for bacterial _____. (20.2)

34. Recombinant DNA plasmids contain _____ from a foreign source. (20.2)

35. PCR is not used to clone functional genes because the resulting DNA copies do not contain _____ genes required for the correct gene expression. (20.2)

36. In PCR, the two strands of the DNA molecule to be copied are separated by gentle _____. (20.2)

37. One heating and cooling cycle in PCR will produce _____ DNA molecules from one original DNA molecule. (20.2)

38. DNA molecules being amplified by PCR will be mixed with _____, _____, and _____ _____. (20.2)

39. Bacteria can manufacture human _____ when they express human genes carried in a recombinant plasmid. (20.2)

## Labeling

*Label the diagram of recombinant DNA production below with the appropriate term or description.*

a. DNAs are mixed. Human fragments line up with plasmid by base pairing of exposed single-strand regions.

b. Plasmids are absorbed by bacteria.

c. Human DNA fragments

d. DNA from bacterial and human cells are isolated.

e. DNA ligase is added, to connect human and plasmid DNA together.

f. Both human DNA and plasmid DNA are cut with the same restriction enzyme.

g. Human DNA containing a gene of interest

h. Bacteria containing the recombinant plasmids of interest are selected and cloned.

i. Bacterial plasmid

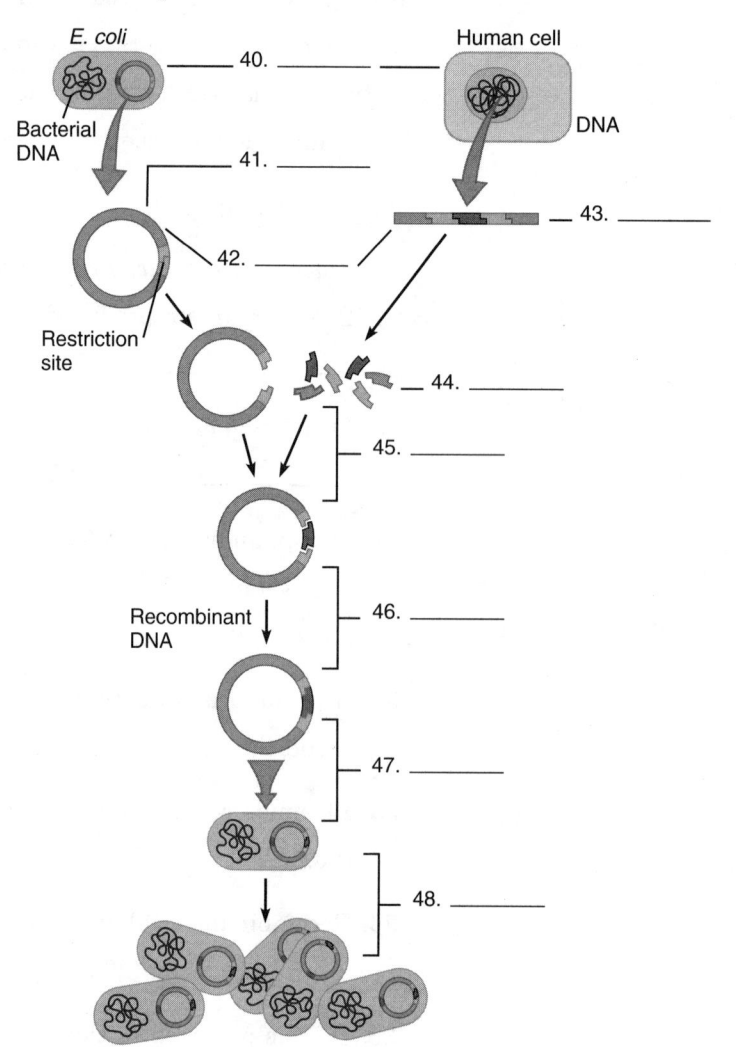

**Figure 20.1**

20.3 Genetic engineering creates transgenic organisms

20.4 Gene therapy: The hope of the future

# Crossword Puzzle

**Across**

6. Microinjections of DNA into a cell is successful less than ____% of the time.
7. Human insulin is currently produced in genetically engineered _____.
11. Cancer research includes the possibility of adding genes for _____ to a patient's cancer cells.
12. Gene therapy is one of the potential benefits of the Human _____ Project.
14. Transgenic animals are _____ to produce.
16. When transgenic plants incorporate human genes they produce human _____.
17. Gene therapy for cystic fibrosis may someday involve the use of _____ spray.
18. The process of producing pharmaceuticals in farm animals is called gene _____.
19. Transgenic bacteria aid the production of _____ when they produce foreign antigens.

**Down**

1. The preferred vector for creating transgenic animals.
2. A genetically engineered vaccine against _____ is currently on the market.
3. The creation of transgenic animals may begin with insertion of foreign DNA into _____ eggs.
4. Viruses used as vectors to create transgenic animals must first be genetically _____.
5. Concerns exist that the use of genetically engineered food crops may lead to some crop _____.
8. _____ organisms carry one or more foreign genes.
9. Something that transports genes into a foreign cell.
10. Gene therapy has successfully treated severe combined _____ deficiency disease.
13. Fatty sphere, which may be able to carry recombinant plasmids into animal cells.
15. Gene _____ involves repairing or replacing damaged human genes.

## Short Answer

20. List three human hormones that are currently being produced by transgenic bacteria.

21. Why are vaccines produced by transgenic bacteria safer than viruses made from weakened or killed disease-causing organisms?

22. What are two of the biggest challenges in gene therapy?

# Chapter Test

## Multiple Choice

1. Which of the following functions as a primer in DNA sequencing?
   a. DNA polymerase
   b. nucleotides labeled with a fluorescent marker
   c. short, single-stranded pieces of DNA that bind to the end of a double-stranded piece of DNA
   d. short, double-stranded pieces of DNA that bind to the end of a single-stranded piece of DNA

2. Gel electrophoresis uses _____ to separate DNA pieces according to their _____.
   a. enzymes, bond structure
   b. primers, base sequence
   c. an electrical field, size
   d. a fluoroscope, altered DNA nucleotides

3. All of the following are associated with DNA sequencing except:
   a. DNA nucleotides labeled with a fluorescent marker.
   b. single-stranded DNA primers.
   c. gel electrophoresis.
   d. plasmids incubated with DNA fragments.

4. Recombinant DNA:
   a. has existed for as long as organisms have carried out sexual reproduction.
   b. results from the selective breeding of plants and animals.
   c. produced from the intentional cutting and splicing of DNA molecules has existed for only a few decades.
   d. all of the above.

5. In nature, plasmids:
   a. are found in bacteria and help to protect the bacteria from viral infection.
   b. are found in viruses and help viral DNA splice itself into a host chromosome.
   c. are found in plants and carry specific genes that are used in plant reproduction.
   d. are found in animals and help to provide protection against bacterial infection.

6. Restriction enzymes cut DNA:
   a. between specific base pairs.
   b. in palindromic sequences.
   c. between hydrogen bonded base pairs.
   d. a and b
   e. a, b, and c

7. Cut fragments of DNA can be bound together by:
   a. DNA polymerase.
   b. DNA ligase.
   c. plasmids.
   d. restriction enzymes.

8. Which of the following is *not* required to create a recombinant plasmid?
   a. DNA polymerase
   b. DNA ligase
   c. foreign DNA
   d. restriction enzymes

9. Place the following steps in the order they would occur during the creation of recombinant plasmids.
   I. addition of DNA ligase
   II. isolate DNA from a foreign source
   III. mix foreign DNA fragments with cut plasmids
   IV. introduce recombinant plasmid into bacteria
   V. cut foreign DNA and plasmids with restriction enzymes
   a. I, II, III, IV, V
   b. IV, I, II, V, III
   c. II, V, III, I, IV
   d. II, III, V, I, IV

10. When recombinant plasmids are created to clone human genes:
    a. retroviruses are required to complete the cloning process.
    b. bacteria containing the desired gene must be identified and isolated prior to cloning.
    c. the plasmids are able to reproduce and clone the genes independently, without bacteria.
    d. the gene of interest must first be microinjected into the bacteria.

11. The polymerase chain reaction makes it possible to:
    a. express all the genes in a particular DNA sequence.
    b. amplify a small amount of DNA.
    c. create a large quantity of DNA polymerase for use in genetic engineering.
    d. create transgenic organisms.

12. During PCR, DNA strands are unwound during _____, and complementary strands are formed during _____.
    a. heating, heating
    b. heating, cooling
    c. cooling, heating
    d. cooling, cooling

13. Transgenic organisms are organisms that:
    a. are produced by selective breeding.
    b. carry recombinant plasmids.
    c. are resistant to diseases.
    d. carry one or more foreign genes.

14. Transgenic bacteria are currently used to produce all of the following except:
    a. beta-carotene.
    b. insulin.
    c. human growth hormone.
    d. erythropoietin.

15. Bacteria that produce vaccines have been engineered to express the gene for:
    a. specific antibodies.
    b. human proteins affected by the disease.
    c. the immune cells normally involved in fighting the disease.
    d. the antigen of the disease-causing organism.

16. Using genetically engineered bacteria to produce vaccines is difficult because:
    a. a specific gene must be located and transferred to the bacteria.
    b. the process is time-consuming and expensive.
    c. microorganisms evolve too rapidly to guarantee that a vaccine will always be effective.
    d. all of the above.

17. Transgenic plants:
    a. can be engineered to express bacterial genes, but not human genes.
    b. have been created, but useful applications have not yet been identified.
    c. do not pose any safety or environmental concerns.
    d. have been engineered for herbicide resistance and vitamin delivery.

18. New cancer research involves introduction, by way of a viral vector, of the mda-7 gene into malignant tumors. As a result, malignant tumor cells:
    a. begin to secrete cell-surface proteins that help the immune system to recognize the malignant cells.
    b. are unable to stimulate the growth of the new blood vessels required to nourish the tumor.
    c. lose the ability to metastasize.
    d. die as a result of apoptosis.

19. Transgenic animals are most often created when:
    a. foreign DNA is microinjected into a fertilized egg.
    b. animal cells readily take up recombinant plasmids.
    c. animals are infected by engineered bacteria.
    d. a single defective gene is replaced by a normal gene.

20. Genetic engineering:
    a. is widely accepted as a valuable tool.
    b. has the potential to provide cures for many human health concerns.
    c. raises many ethical problems.
    d. all of the above.

21. The ability to repair or replace damaged human genes is an example of:
    a. gene therapy.
    b. recombinant DNA technology.
    c. genetic engineering.
    d. all of the above.

22. Which of the following is a false statement about gene therapy?
    a. Gene therapy has the potential to help in the treatment of many diseases.
    b. One of the challenges of gene therapy is finding a way to deliver new DNA to all necessary body cells.
    c. Diseases that are successfully treated with gene therapy cannot be inherited by an affected individual's offspring.
    d. Retroviruses are potential vectors for use in gene therapy.

23. The first disease to be treated successfully with gene therapy was:
    a. cystic fibrosis.
    b. SCID.
    c. cancer.
    d. diabetes.

24. Retroviruses have the potential to be effective vectors in delivering DNA to human cells because:
    a. they do not normally infect human cells.
    b. they never mutate.
    c. they are easily microinjected.
    d. they can splice their genetic information into the DNA of human cells.

25. While retroviruses are capable of introducing foreign DNA into a human genome:
    a. this has never been done successfully.
    b. they may disrupt important regulatory genes when they insert foreign DNA at random locations in a chromosome.
    c. their use is limited because they only insert foreign DNA during G1 of the host cell cycle.
    d. the difficulty of creating a recombinant virus makes their common use as a vector unlikely.

# Key Concept Review Questions

*Each of the Key Concepts listed at the beginning of the study guide has been rewritten as a question below. After successfully completing the study guide exercises and the Chapter Test, you should be able to answer each of these questions. Refer to the Key Concept List at the beginning of this chapter to check your answers.*

1. What is biotechnology?
2. What does DNA sequencing allow scientists to determine?
3. What structures, enzymes, and processes are required for gene sequencing?

4. What is recombinant DNA? What three processes are involved in recombinant DNA technology?
5. List four things needed to produce recombinant bacteria.
6. What effect do restriction enzymes have on DNA?
7. What is the action of DNA ligase?
8. What are plasmids and where are they found in nature?
9. What can plasmids carry into bacteria? What must bacteria contain to manufacture a foreign protein?
10. What is PCR?
11. Briefly describe the process of PCR.
12. What are transgenic organisms?
13. What are transgenic bacteria and what can they produce?
14. What are transgenic plants? List some examples of the benefits of transgenic plants.
15. What are some of the concerns associated with the use of transgenic plants?
16. Why does research on transgenic animals continue even though they are difficult to create?
17. What is one technique used to create transgenic animals?
18. What is gene therapy?
19. How are retroviruses used in gene therapy?
20. What disease has been successfully treated with gene therapy? List two diseases that are currently being investigated for treatment with gene therapy.

# Answer Key

## Sections 20.1, 20.2

**1.**f; **2.**i; **3.**j; **4.**c; **5.**h; **6.**e; **7.**b; **8.**k; **9.**a; **10.**g; **11.**d; **12.** base; **13.** test tube; **14.** primers; **15.** strands; **16.** synthesis; **17.** labeled; **18.** fluorescent; **19.** synthesis; **20.** polymerase; **21.** lengths; **22.** labeled; **23.** gel electrophoresis; **24.** size; **25.** smaller; **26.** larger; **27.** laser; **28.** complementary; **29.** recombinant; **30.** base pairs; **31.** GACGTC; **32.** bacteria; **33.** reproduction; **34.** DNA; **35.** regulatory; **36.** heating; **37.** two; **38.** primers, nucleotides, DNA polymerase; **39.** proteins; **40.**d; **41.**i; **42.**f; **43.**g; **44.**c; **45.**a; **46.**e; **47.**b; **48.**h

## Sections 20.3, 20.4

**Crossword Puzzle: 1.** retroviruses; **2.** hepatitis B; **3.** fertilized; **4.** engineered; **5.** failures; **6.** ten; **7.** bacteria; **8.** transgenic; **9.** vector; **10.** immuno; **11.** interleukins; **12.** Genome; **13.** liposomes; **14.** difficult;

**15.** therapy; **16.** proteins; **17.** nasal; **18.** pharming; **19.** vaccines; **20.** insulin, human growth hormone, erythropoietin; **21.** Viruses made from weakened or killed disease-causing organisms can sometimes cause the disease, while viruses produced in recombinant bacteria contain only the antigen of the disease-causing organism and so carry no risk of causing disease; **22.** inserting recombinant DNA into a sufficient number of cells to correct a defect, and treating a disease without being able to prevent the passing of the disease to offspring

## Chapter Test

**1.**c; **2.**c; **3.**d; **4.**d; **5.**a; **6.**d; **7.**b; **8.**a; **9.**c; **10.**b; **11.**b; **12.**b; **13.**d; **14.**a; **15.**d; **16.**d; **17.**d; **18.**d; **19.**a; **20.**d; **21.**d; **22.**c; **23.**b; **24.**d; **25.**b

# 21

# Development and Aging

## Chapter Summary and Key Concepts

*After reading and studying this chapter you should know the following:*

### Sections 21.1, 21.2

1. Sperm travel through the vagina, cervix, uterus, and oviduct in an attempt to fertilize the egg.

2. The egg is released as an immature secondary oocyte and completes meiosis II only after fertilization occurs.

3. Fertilization occurs in the upper third of the oviduct when the nuclei of the sperm and ovum fuse to form a zygote.

4. Twins may be fraternal, identical, or conjoined.

5. The four processes of development are: a series of cell divisions called *cleavage,* a process of physical change called *morphogenesis,* the development of specialized form and function in cells called *differentiation,* and an increase in size called *growth.*

### Sections 21.3, 21.4

6. Prenatal development is divided into three stages: pre-embryonic, embryonic, and fetal development.

7. In pre-embryonic development, the zygote becomes a morula and then a blastocyst that implants in the endometrium.

8. During embryonic development the embryo consist of three germ layers called the endoderm, the mesoderm, and the ectoderm.

9. Four extraembryonic membranes form during embryonic development to support the embryo: the amnion, the allantois, the yolk sac, and the chorion.

10. The placenta consists of a maternal portion, the endometrium, and an embryonic portion, the chorion and chorionic villi.

11. The umbilical cord connects the placenta to the embryo's circulation.

12. By the end of embryonic development the embryo is distinctly human in appearance.

### Sections 21.5, 21.6

13. During fetal development, growth is rapid, the fetus begins to move, surfactant is produced, and life outside the womb becomes possible.

14. Birth involves three stages of labor and delivery: dilation, expulsion, and afterbirth.

15. At birth, changes in the fetal circulation sustain breathing by the newborn.

16. Lactation delivers first colostrum, and then milk, to the newborn.

### Sections 21.7, 21.8, 21.9

17. The stages of human development after birth are the neonatal period, infancy, childhood, adolescence, and adulthood.

18. Aging is a process of deterioration that occurs over time.

19. Aging is marked by changes in major body systems.

20. Death is the irreversible cessation of processes required to sustain life.

# Exercises

*Complete the exercises for each section after you have read and studied the section. If you cannot answer some questions, or answer them incorrectly, return to the chapter and review this information. You may find it helpful to work on only one section at a time. When you have completed all sections, take the Chapter Test as an indicator of your mastery of this topic.*

**21.1 Fertilization occurs when sperm and egg unite**

**21.2 Development: Cleavage, morphogenesis, differentiation, and growth**

## Labeling and Short Answer

Use the terms below to label Figure 21.1 of a sperm and a secondary oocyte, and then answer the questions.

head        nucleus           corona radiata
cytoplasm   midpiece          first polar body
acrosome    plasma membrane   zona pellucida
tail

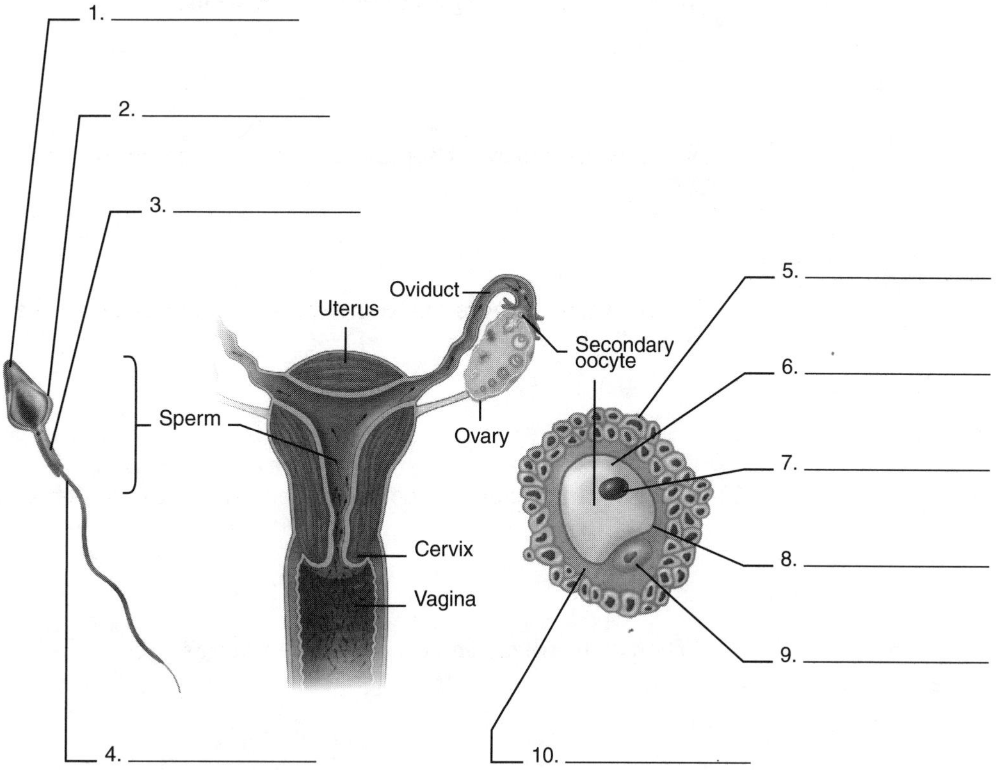

1. _____
2. _____
3. _____
4. _____
5. _____
6. _____
7. _____
8. _____
9. _____
10. _____

**Figure 21.1**

11. What part of a sperm contains enzymes that aid the sperm's entry into an egg?

12. Where does fertilization usually occur?

13. List, in order, the structures in the female through which the sperm must pass before they reach the egg:

14. In the absence of fertilization, how long does the egg survive?

15. What layers of the secondary oocyte must the sperm penetrate to reach the cell membrane?

16. What marks the beginning of fertilization?

17. What events prevent fertilization of the egg by multiple sperm?

18. What event triggers the completion of meiosis II by the secondary oocyte?

19. What is the secondary oocyte called when it completes meiosis II?

20. When is fertilization considered complete?

21. What structures are lost as the ovum becomes a zygote?

## Matching

\_\_\_\_ 22. **twins**  a. cell divisions without cell growth that occur in the first 4 days after fertilization

\_\_\_\_ 23. **fraternal twins**  b. an increase in size that begins with implantation

\_\_\_\_ 24. **identical twins**  c. the process of physical changes that occur during development

\_\_\_\_ 25. **conjoined twins**  d. twins that arise from the ovulation of more than one oocyte

___ 26. **cleavage**  e. the most common form of multiple births

___ 27. **morphogenesis**  f. the development of specialized forms and functions in individual cells

___ 28. **differentiation**  g. twins that arise when cells begin, but do not complete separation—twins remain attached to each other

___ 29. **growth**  h. twins that arise from a single zygote when cells break apart

**21.3  Pre-embryonic development: The first 2 weeks**

**21.4  Embryonic development: Weeks 3 to 8**

## Labeling

*Use the terms below to label each structure in Figure 21.2.*

a. pre-embryo
b. trophoblast
c. endoderm
d. diploid nucleus
e. yolk sac
f. ectoderm
g. inner cell mass
h. morula
i. trophoblast cells
j. zygote
k. embryonic disc
l. blastocyst
m. amniotic cavity
n. embryo
o. hollow cavity
p. endoderm
q. trophoblast cells
r. amniotic cavity
s. embryonic disc
t. mesoderm
u. ectoderm

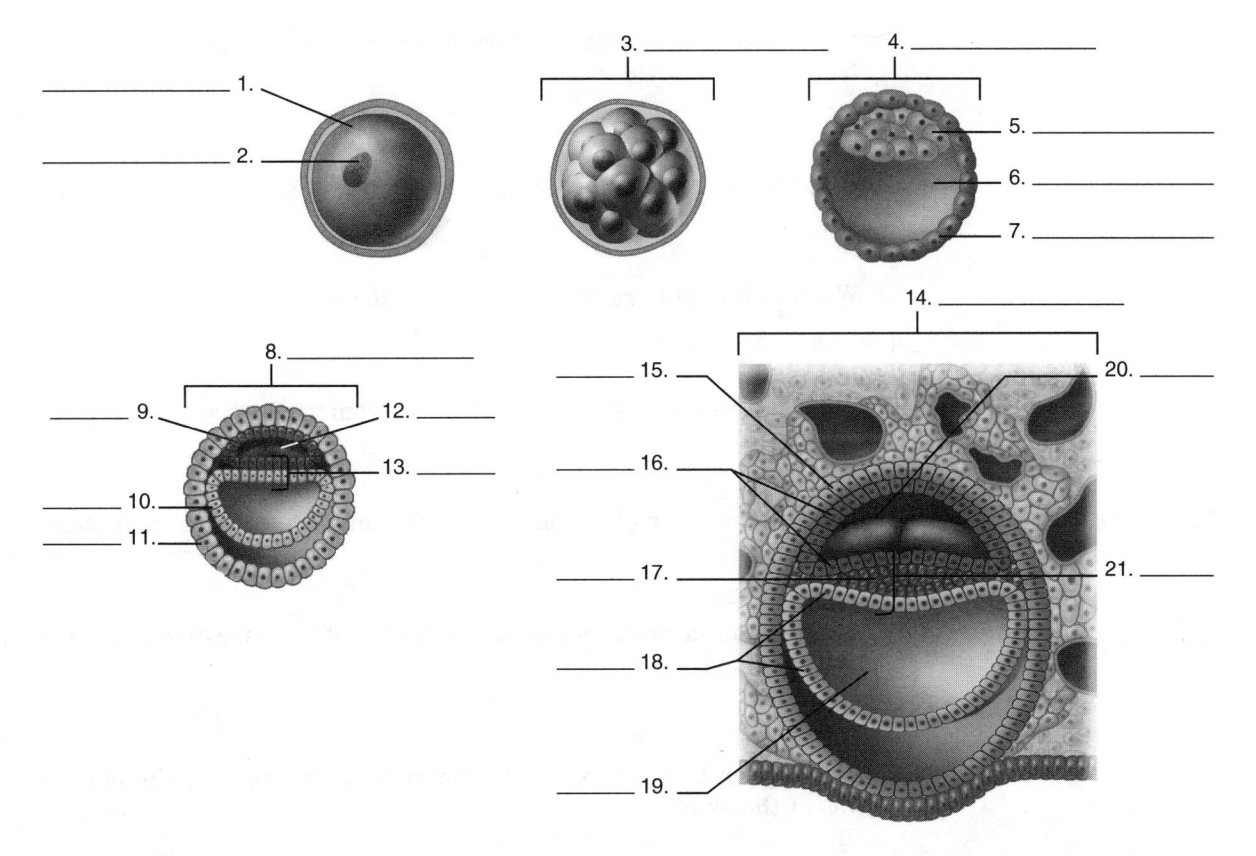

**Figure 21.2**

## Short Answer

*Refer to Figure 21.2 as you answer the following questions.*

22. What time period after conception is considered to be the pre-embryonic stage?

23. Which structure in Figure 21.2 leaves the oviduct and enters the uterus?

24. Which structure in Figure 21.2 is the first to undergo differentiation and morphogenesis?

25. Which part of the blastocyst will eventually become the embryo?

26. Which structure in Figure 21.2 burrows into the endometrium of the uterus?

27. What is this burrowing process called?

28. What part of the blastocyst secretes enzymes that digest endometrial cells?

29. What marks the end of the pre-embryonic stage?

30. When is the fetus referred to as an embryo?

31. What marks the beginning of the embryonic stage?

32. Which germ layer gives rise to muscle, connective tissue, and bone?

33. Which germ layer gives rise to the lining of the digestive tract and several glands?

34. Which germ layer gives rise to the nervous system and the epidermis of the skin?

## Labeling

Use the terms below to label Figure 21.3 below.

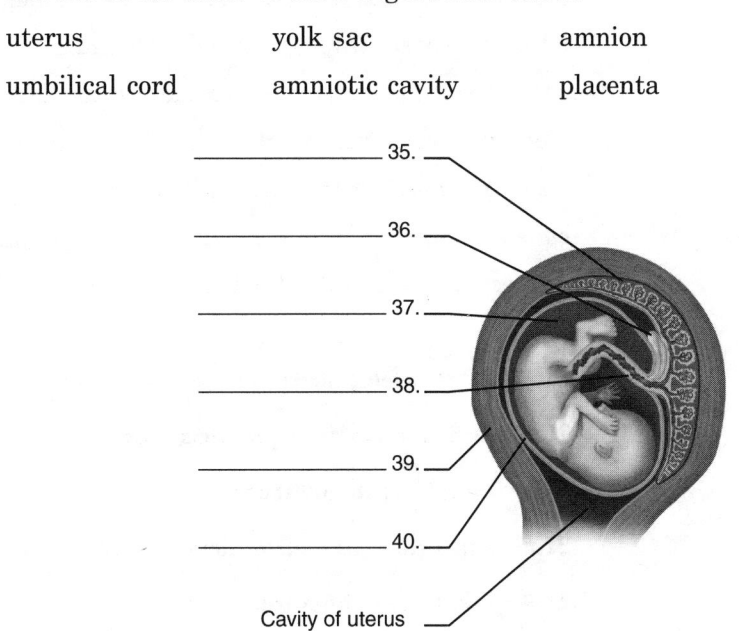

uterus   yolk sac   amnion
umbilical cord   amniotic cavity   placenta

_____ 35.
_____ 36.
_____ 37.
_____ 38.
_____ 39.
_____ 40.

Cavity of uterus

**Figure 21.3**

## Completion

Use the terms below to complete the paragraph, and then repeat this exercise with the terms covered.

| trophoblast | amnion | gonads | chorion |
| endometrium | digestive | yolk sac | umbilical cord |
| allantois | chorion | embryonic | amnion |
| endometrium | placenta | villi | umbilical cord |

Extra-embryonic membranes and the placenta develop early in the (41) _____ stage. The four extra-embryonic membranes are the (42) _____, (43) _____, (44) _____, and the (45) _____. The (46) _____ is also known as the "bag of waters" and is filled with fluid. The allantois is a temporary membrane that helps form the blood vessels of the (47) _____. The yolk sac becomes part of the fetal (48) _____ tract and provides the germ cells that will move into the (49) _____ and eventually give rise to eggs and sperm. The chorion develops primarily from (50) _____, and will form important structures for gas and nutrient exchange in the (51) _____.

The placenta forms from the (52) _____ of the fetus and the (53) _____ of the mother, and connects to the fetal circulation through the (54) _____. Chorionic cells secrete enzymes that damage the (55) _____ and cause bleeding, providing the medium for gas and nutrient exchange and waste removal for the fetus, while fingerlike projections of the chorion called chorionic (56) _____ contain small capillaries that connect to the umbilical arteries and veins.

**21.5  Fetal development: Nine weeks to birth**

**21.6  Birth and the early postnatal period**

**21.7  From birth to adulthood**

**21.8  Aging takes place over time**

**21.9  Death is the final transition**

## Short Answer

*Write the main events that occur in each month of fetal development below. There are no answers for this exercise in the answer key. Refer to Section 21.5 in the textbook to check your answers.*

1. Third month:

2. Fourth month:

3. Fifth month:

4. Sixth month:

5. Seventh through ninth months:

## Fill-in-the-Blank

*Referenced sections are in parentheses.*

6. Labor is triggered by maturation of the fetal _____ _____. (21.6)

7. _____ hormones secreted by the fetus stimulate the placenta to increase production of _____. (21.6)

8. Contraction of the uterus during labor is a _____ feedback cycle. (21.6)

9. The three phases of labor and delivery are _____, _____, and _____. (21.6)

10. Stage 1 of labor and delivery includes the breaking of the _____ and usually lasts _____ hours. (21.6)

11. Stage 2 of labor and delivery lasts from full _____ _____ to delivery. (21.6)

12. A(n) _____ is a surgical incision that enlarges the vaginal opening for birth. (21.6)

13. During stage 3 of labor and delivery, strong contractions after delivery of the infant expel the _____ _____ and the _____. (21.6)

14. Pulmonary _____ produced by the infant reduces surface tension in the _____. (21.6)

15. In the fetus, blood flows from the umbilical vein to the inferior vena cava through the _____ _____. (21.6)

16. In the fetal circulation, blood flows from the right atrium to the left atrium through the _____ _____. (21.6)

17. After birth, the umbilical blood vessels regress to become _____ tissue. (21.6)

18. During the first few days after birth, the mother produces a watery milk called _____. (21.6)

19. In the mother, the homone _____ stimulates milk production and the hormone _____ stimulates delivery of milk. (21.6)

20. The first month of life is referred to as the _____ period. (21.7)

21. During the first 15 months of life, most infants will _____ their weight. (21.7)

22. During infancy, the _____ system is the slowest to develop. (21.7)

23. The brain grows to 95% of its final size during _____. (21.7)

24. In adolescence, puberty is triggered when the hypothalamus begins secreting _____. (21.7)

25. During the period of _____, a lumbar curve develops and abdominal muscles become stronger. (21.7)

26. A section of disposable DNA found on the end of every DNA strand and believed to play a role in aging is called a _____. (21.8)

27. Menopause occurs when the ovaries lose their responsiveness to _____ and _____. (21.8)

28. Describe in one or two words the changes that occur in each of the following with age: (21.8)
    a. lung tissue _____
    b. heart walls _____
    c. bone mass _____
    d. muscle mass _____
    e. lens of the eye _____
    f. viable sperm produced _____
    g. kidney mass _____
    h. prostate gland _____

29. Death is defined as the _____ cessation of circulatory and respiratory, or brain, function. (21.9)

# Chapter Test

## Multiple Choice

1. In the female reproductive tract, sperm can survive for _____ and the egg survives for _____.
   a. one day, one day
   b. several days, one day
   c. one day, several days
   d. several days, several days

2. The protective layer surrounding the secondary oocyte is the:
   a. follicle.
   b. granulosa.
   c. corona radiata.
   d. zona pellucida.

3. Fertilization is considered complete when:
   a. the sperm enters the egg.
   b. the zona pellucida breaks down.
   c. the nuclei of the sperm and ovum unite.
   d. the sperm attaches to the cell membrane of the secondary oocyte.

4. Ovulation of more than one oocyte in a cycle may result in:
   a. fraternal twins.
   b. maternal twins.
   c. conjoined twins.
   d. identical twins.

5. The primary cause of morphogenesis is:
   a. cleavage.
   b. differentiation.
   c. fertilization.
   d. growth.

6. Which of the following is true of the blastocyst?
   a. It develops during the embryonic stage.
   b. It is a ball of 32 identical cells.
   c. It consists of a trophoblast and an inner cell mass.
   d. It develops into a morula.

7. Which of the following does not occur during cleavage?
   a. a series of cell divisions
   b. cell differentiation
   c. cell growth
   d. b and c
   e. a, b, and c all occur during cleavage

8. Implantation in the endometrium occurs when:
   a. trophoblast cells secrete enzymes that digest endometrial cells.
   b. contact of the blastocyst with the endometrium initiates the growth of new blood vessels.
   c. the morula leaves the oviduct and enters the uterus.
   d. fluid from the amniotic cavity contacts endometrial cells and triggers burrowing of the blastocyst.

9. Muscle and connective tissue arise from which germ layer?
   a. endoderm
   b. ectoderm
   c. mesoderm
   d. trophoderm

10. The placenta is formed during embryonic development from the:
    a. chorion and endometrium.
    b. amnion and chorion.
    c. amnion and endometrium.
    d. chorion and myometrium.

11. As an endocrine organ, the placenta secretes:
    a. hCG after the corpus luteum degenerates.
    b. progesterone and estrogen to maintain and promote pregnancy when the corpus luteum degenerates.
    c. LH and FSH to stimulate follicle development in the ovaries.
    d. progesterone to maintain the corpus luteum.

12. The neural groove that forms during embryonic development will become:
    a. muscle.
    b. bone.
    c. brain and spinal cord.
    d. heart and blood vessels.

13. It is difficult to distinguish a human embryo from any other vertebrate embryo prior to the _____ week of development.
    a. fourth.
    b. sixth.
    c. eighth.
    d. 12th.

14. Male and female gonads develop in the:
    a. first trimester.
    b. second trimester.
    c. third trimester.
    d. just before birth.

15. A developing human is called a fetus:
    a. after the pre-embryonic stage.
    b. at the beginning of the second trimester.
    c. after the eighth week.
    d. in the third trimester.

16. Contractions of the uterus during labor:
    a. increase the release of oxytocin.
    b. decrease the release of oxytocin.
    c. increase the release of prolactin.
    d. increase the release of ADH.

17. The hormone _____ stimulates milk production in the mother.
    a. oxytocin
    b. estrogen
    c. progesterone
    d. prolactin

18. Surfactant is important in:
    a. initiating uterine contractions prior to birth.
    b. preventing blood loss when the umbilical cord is severed.
    c. redirecting fetal circulatory pathways in the newborn.
    d. reducing surface tension in the alveoli of the lungs.

19. A shunt that directs fetal circulation from the pulmonary artery to the aorta is the:
    a. ductus arteriosus.
    b. ductus venosus.
    c. foramen ovale.
    d. pulmonary stenosis.

20. Colostrum:
    a. is a pasty substance that covers the newborn's skin.
    b. is a watery milk produced by the breast.
    c. is a fluid lost with the afterbirth.
    d. is an enzyme produced by the myometrium during labor.

21. The period of life from age 2–15 months is termed:
    a. neonatal.
    b. childhood.
    c. infancy.
    d. adolescence.

22. The _____ system lags behind most other systems in its development.
    a. cardiovascular
    b. immune
    c. muscular
    d. nervous

23. All of the following are accepted theories of the cause of aging except:
    a. genetic programming determines the timing of cell death.
    b. environmental toxins have led to a shortened life span.
    c. accumulated cell damage interferes with cell repair.
    d. interdependence of body systems promotes aging when one system begins to fail.

24. The finding that severe caloric restriction in some animals prolongs life supports the theory that aging is due to:
    a. accumulated damage to DNA and errors in DNA replication.
    b. an internal genetic counting mechanism, possibly involving telomeres, that limits the number of cell divisions in each cell.
    c. the decline in one body system that affects other body systems.
    d. poor nutrition.

25. Death may be defined as a cessation of either circulatory or respiratory function or of brain function. In either definition, a key word preceding "cessation" is:
    a. complete.
    b. prolonged.
    c. irreversible.
    d. acute.

# Key Concept Review Questions

*Each of the Key Concepts listed at the beginning of this chapter has been rewritten as a question below. After successfully completing the study guide exercises and the Chapter Test, you should be able to answer each of these questions. Refer to the Key Concepts list at the beginning of this chapter to check your answers.*

1. Trace the path of a sperm from the time it enters the female reproductive tract until it reaches the egg.

2. When does the secondary oocyte complete meiosis II?

3. Where does fertilization usually occur? What is formed by the fusion of the nuclei of the sperm and the ovum?

4. What are the three types of twins?

5. What are the four processes involved in development?

6. What are the three stages of prenatal development?

7. What are the developmental structures that occur in the pre-embryonic stage? Which one becomes implanted in the endometrium?

8. What are the three germ layers that develop during the embryonic stage?

9. Name the four extra-embryonic membranes that develop during the embryonic stage.

10. What structures make up the placenta?

11. What two structures are connected by the umbilical cord?

12. The embryo is distinctly human in appearance by the end of what prenatal stage?

13. What are some of the events that occur during fetal development?

14. What are the three stages of labor and delivery?

15. Is the fetal circulatory system capable of sustaining breathing in the newborn?

16. What substances are provided to the newborn during lactation?

17. What are the five stages of human development that occur after birth?

18. What is aging?

19. Aging is marked by changes in what parts of the body?

20. What is the definition of death?

# Answer Key

## Sections 21.1, 21.2

**1.** acrosome; **2.** head; **3.** midpiece; **4.** tail; **5.** corona radiate; **6.** cytoplasm; **7.** nucleus; **8.** plasma membrane **9.** first polar body; **10.** zona pellucida; **11.** acrosome; **12.** upper third of the oviduct; **13.** vagina, cervix, uterus, oviduct; **14.** 24 hours; **15.** zona pellucida and corona radiate; **16.** entry of the sperm's nucleus into the egg; **17.** sperm entry triggers the release of enzymes that cause changes in the zona pellucida making it impenetrable to additional sperm; **18.** entry of the sperm; **19.** zygote; **20.** when the nuclei of the sperm and mature ovum unite; **21.** follicle cells and polar bodies; **22.**e; **23.**d; **24.**h; **25.**g; **26.**a; **27.**c; **28.**f; **29.**b

## Sections 21.3, 21.4

**1.** j; **2.** d; **3.** h; **4.** l; **5.** g; **6.** o; **7.** b; **8.** a; **9.** f; **10.** c; **11.** i; **12.** m; **13.** k; **14.** n; **15.** q; **16.** u; **17.** t; **18.** p; **19.** e; **20.** r; **21.** s; **22.** 2 weeks; **23.** morula; **24.** morula; **25.** embryonic disc; **26.** blastocyst; **27.** implantation; **28.** trophoblast cells; **29.** the appearance of an amniotic cavity, and of ectoderm and endoderm in the embryonic disc; **30.** week 3 through week 8; **31.** the appearance of mesoderm; **32.** mesoderm; **33.** endoderm; **34.** ectoderm; **35.** placenta; **36.** yolk sac; **37.** amniotic cavity; **38.** umbilical cord; **39.** uterus; **40.** amnion; **41.** embryonic; **42.** amnion; **43.** allantois; **44.** yolk sac; **45.** chorion; **46.** amnion; **47.** umbilical cord; **48.** digestive; **49.** gonads; **50.** trophoblast; **51.** placenta; **52.** chroion; **53.** endometrium; **54.** umbilical cord; **55.** endometrium; **56.** villi

## Sections 21.5, 21.6, 21.7, 21.8, 21.9

**1–5:** See textbook; **6.** pituitary gland; **7.** Steroid, estrogen; **8.** positive; **9.** dilation, expulsion, afterbirth; **10.** water, 6–12; **11.** cervical dilation; **12.** episiotomy; **13.** umbilical cord, placenta; **14.** surfactants, lungs; **15.** ductus venosus; **16.** foramen ovale; **17.** connective; **18.** colostrum; **19.** prolactin, oxytocin; **20.** neonatal; **21.** triple; **22.** immune; **23.** childhood; **24.** GnRH; **25.** childhood; **26.** telomere; **27.** LH, FSH; **28.** a. less elastic, b. stiffen, c. decreases, d. decreases, e. stiffens, f. decreases, g. decreases, h. enlarges; **29.** irreversible

## Chapter Test

**1.** b; **2.** d; **3.** c; **4.** a; **5.** b; **6.** c; **7.** d; **8.** a; **9.** c; **10.** a; **11.** b; **12.** c; **13.** a; **14.** a; **15.** c; **16.** a; **17.** d; **18.** d; **19.** a; **20.** b; **21.** c; **22.** b; **23.** b; **24.** a; **25.** c

# 22

# Evolution and the Origins of Life

## Chapter Summary and Key Concepts

*After reading and studying this chapter you should know the following:*

### Sections 22.1, 22.2

1. Evolution is the unpredictable and natural process of descent with modification over time.

2. Microevolution occurs as a result of genetic changes.

3. Macroevolution involves large-scale trends in evolution among groups of species.

4. Charles Darwin proposed the theory of descent with modification in the 1800s.

5. Evidence for evolution comes from the fossil record, comparative anatomy and embryology, biochemistry, and biogeography.

6. Fossils are preserved remnants of organisms.

7. The fossil record is incomplete because fossil formation depends on the type of tissue and the environmental conditions.

8. Homologous structures are similar because they evolved from a common ancestor; analogous structures are similar because they evolved for a similar function.

9. Vestigial structures serve little or no function and may be the remains of an ancestral structure.

10. Similarities in vertebrate embryos suggest that all vertebrates evolved from a common ancestor.

11. Similarities in biochemical molecules provide information about how recently species may have diverged.

12. Biogeography is the study of the distribution of plants and animals around the world.

13. Geographic barriers promote evolution.

14. In the process of natural selection, organisms with traits that make them more fit for the environment are able to survive and reproduce.

15. Evolution is the result of the presence of mutations and the process of natural selection.

16. Genetic drift and gene flow can change populations by changing allele frequencies.

**Sections 22.3, 22.4, 22.5**

17. Life on earth began with the formation of organic molecules that became enclosed in lipid-protein membranes.

18. The appearance of photosynthetic organisms allowed oxygen to accumulate in the atmosphere, making possible the eventual appearance of animals.

**Section 22.6**

19. Humans are classified as *Homo sapiens*.

20. The evolution of *Homo sapiens* required ancestral species to walk upright and develop a larger brain.

# Exercises

*Complete the exercises for each section after you have read and studied the section. If you cannot answer some questions, or answer them incorrectly, return to the chapter and review this information. You may find it helpful to work on only one section at a time. When you have completed all sections, take the Chapter Test as an indicator of your mastery of this topic.*

22.1  **Evidence for evolution comes from many sources**

22.2  **Natural selection contributes to evolution**

# Crossword Puzzle

**Across**

4. The complete disappearance of a species
5. Genetic _____ results in random changes in allele frequency in a population.
8. Gene _____ refers to the effect of migration on the allele frequency in a population.
9. Individuals of the same species in a particular area

**Down**

1. Natural _____ is a process where the most fit individuals survive and reproduce.
2. Results from genetic changes that give rise to new species
3. Body structures that are similar due to a common ancestor
6. Adaptive _____ occurs when new species develop from a single ancestor in a short time.
7. Body structures with a similar function but not a similar structure
8. Preserved remnant of an organism

# Fill-in-the-Blank

*Referenced sections are in parentheses.*

10. The three components of evolution are _____ _____ _____, _____ _____, and that evolution is _____ and _____. (Intro)

11. Charles Darwin's phrase for evolution was _____ _____ _____. (Intro)

12. Darwin asserted that life arose _____ time(s). (Intro)

13. Hard tissues are preserved as fossils only when they are quickly covered with _____ or _____ _____. (22.1)

14. _____ involves large-scale evolutionary trends or changes that apply to whole groups of species. (Intro)

15. Sedimentary deposits are laid layer upon layer in a process called _____. (22.1)

16. When fossils exist in sedimentary rock, the oldest fossils will be located in the _____ layers of rock. (22.1)

17. Radiometric dating in fossils older than 50,000 years requires the use of isotopes of _____. (22.1)

18. In humans, the muscles that wiggle the ears are examples of _____ _____. (22.1)

19. Comparative biochemistry suggest a close relationship between species when they have similar or identical _____ _____. (22.1)

20. One species becomes two different species when evolutionary paths _____. (22.1)

21. _____ is the study of the distribution of plants and animals around the world. (22.1)

22. Two hundred million years ago all continents were joined in one land mass called _____. (22.1)

23. Adaptive radiation often occurs after a _____ _____. (22.2)

24. The _____ _____ alters the gene pool of a population when a major catastrophe wipes out most of the population. (22.2)

## Short Answer

*Refer to Figure 22.1 below to answer the questions.*

25. In figure (a):

    a. The human arm and the bird's wing are _____ structures.

    b. Why is there similarity in the structure of the arm and the wing?

    c. What might explain the differences in these structures?

26. In figure (b):

    a. The bird wing and the insect wing are examples of _____ structures.

    b. Did these two wings evolve from a common structure in a common ancestor?

    c. What might explain the similarities between the two wings?

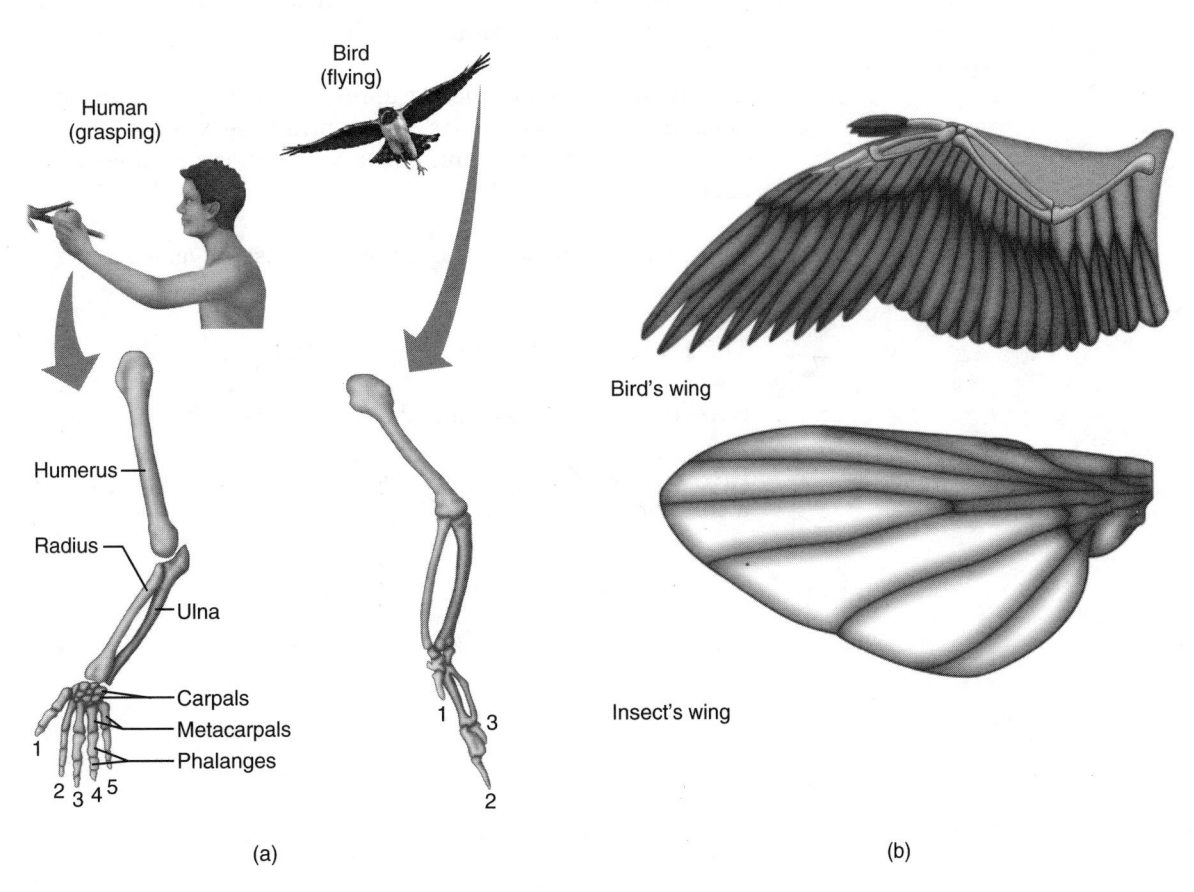

**Figure 22.1**

**22.3** The young earth was too hot for life

**22.4** The first cells were able to live without oxygen

**22.5** Photosynthetic organisms altered the course of evolution

## Ordering

*Label each of the following events with the letters a–k to indicate the sequence in which they occurred as life developed on Earth.*

_____ 1. Heat vaporized all water, preventing the formation of oceans.

_____ 2. Oxygen, a by-product of photosynthesis, begins to accumulate in the atmosphere.

_____ 3. Self-replicating RNA molecules form on clay templates along the ocean edge.

_____ 4. Multicellular organisms appear.

_____ 5. Mutation allows some cells to develop the capacity for photosynthesis.

_____ 6. The Earth formed about 4.6 billion years ago with a liquid core of metal and a thin outer crust.

_____ 7. Simple organic molecules of amino acids, sugars, and fatty acids form from molecules in the primitive atmosphere, using heat, radiation, and electrical discharges as energy sources.

_____ 8. New cells evolve with the capacity to use oxygen for aerobic metabolism.

_____ 9. The Earth cools enough for oceans to form.

_____ 10. Simple organic compounds dissolve in the sea.

_____ 11. RNA and small organic molecules become enclosed within a lipid-protein membrane.

**Section 22.6 Modern humans came from Africa**

# Completion

*Fill in the table below by writing the scientific name of each taxonomic category, as it applies to humans. Write one or more characteristics that qualify humans for each category.*

| Taxonomic Category | Scientific Name | Human Characteristics |
|---|---|---|
| Kingdom | 1. | 2. |
| Phylum | 3. | 4. |
| Class | 5. | 6. |
| Order | 7. | 8. |
| Family | 9. | 10. |
| Genus | 11. | 12. |
| Species | 13. | 14. |

## Matching

*For each of the characteristics listed below, indicate the letter of the species being described.*

a. ***Homo erectus***
b. ***Australopithecus afarensis***
c. ***Homo sapiens***
d. ***Homo habilis***

_____ 15. first species believed to have migrated out of Africa

_____ 16. the first distinctly human ancestor

_____ 17. modern humans

_____ 18. the only surviving *Homo* species

_____ 19. the first human ancestor to walk upright

_____ 20. lived about 1.8 million years ago

_____ 21. believed to be the earliest direct ancestor of modern humans

_____ 22. believed to have been the first toolmakers

_____ 23. the society of this species probably included hunting and gathering groups that shared food

_____ 24. lived about 3.2 million years ago

_____ 25. lived about 2.4 million years ago

_____ 26. may have evolved from a single colony of *Homo heidelbergensis*

_____ 27. probably the first to have a diet that included meat

_____ 28. arose about 140,000 years ago

_____ 29. was a contemporary of modern humans

_____ 30. showed the greatest degree of sexual dimorphism

# Chapter Test

## Multiple Choice

1. Evolution that occurs as a result of genetic changes that gives rise to a new species is referred to as:
   a. microevolution.
   b. macroevolution.
   c. genetic evolution.
   d. allelic evolution.

2. Which of the following is most likely to form a fossil?
   a. a soft tissue organism buried by volcanic ash
   b. an aquatic organism
   c. a vertebrate organism covered by sediment
   d. an insect ingested by a reptile

3. The fossil record is incomplete because:
   a. we have no fossil record of many species that lacked hard tissues.
   b. it is difficult to accurately date some fossils.
   c. many people do not accept the evidence for evolution found in the fossil record.
   d. many fossils may be difficult to find.
   e. a and d

4. The forelimbs of all vertebrates are homologous. This means:
   a. vertebrates with similar forelimbs evolved in similar environments and experienced the same natural selection processes.
   b. the forelimbs of all vertebrates have identical structures.
   c. all vertebrates descended from a common ancestor.
   d. vertebrate evolution occurred more quickly than that of other organisms.

5. Radioactive dating of fossils younger than 50,000 years uses _____ isotopes, while dating of older fossils uses _____ isotopes.
   a. oxygen, carbon
   b. carbon, potassium
   c. potassium, magnesium
   d. potassium, carbon

6. The coccyx, or tailbone, of a human is a(n) _____ structure.
   a. homologous
   b. vestigial
   c. analogous
   d. divergent

7. The presence of identical or nearly identical biochemical molecules in two species indicates:
   a. recent divergence.
   b. the presence of analogous structures.
   c. that they developed in the same geographical region.
   d. an increased rate of mutation.

8. The bottleneck effect results in the survival of individuals who:
   a. carry genes representing the most fit genes in the population.
   b. are the best adapted for survival and reproduction.
   c. share a very small gene pool that was selected with no regard to fitness.
   d. left the original population and began a new population in a different location.
   e. a and b

9. The effect of the continental drift has been to:
   a. increase the rate of mutation.
   b. decrease the rate of mutation.
   c. isolate related groups of organisms from each other.
   d. maintain Pangaea.

10. A population is a group of individuals:
    a. who share the same alleles.
    b. who occupy the same area.
    c. of the same species who occupy the same area.
    d. of the same species, anywhere on Earth.

11. Random changes in the allele frequency because of chance effects is called:
    a. gene flow.
    b. natural selection.
    c. genetic drift.
    d. genetic change.

12. The largest mass extinction was the:
    a. Triassic.
    b. Jurassic.
    c. Cretaceous.
    d. Diluvian.

13. Short bursts of evolutionary activity are called:
    a. divergence.
    b. adaptive radiation.
    c. natural selection.
    d. selective radiation.

14. The Earth formed about:
    a. 6 billion years ago.
    b. 4.6 billion years ago.
    c. 4 million years ago.
    d. 2 million year ago.

15. Which of the following was not found in the primitive atmosphere of the earth?
    a. $CO_2$
    b. $H_2O$
    c. $O_2$
    d. $N_2$

16. Before life evolved, simple organic molecules formed from:
    a. the oceans.
    b. atmospheric gases.
    c. the rain.
    d. DNA and RNA.

17. The first self-replicating molecule was probably:
    a. DNA.
    b. RNA.
    c. lipids.
    d. proteins.

18. The evolution of photosynthetic organisms resulted in:
    a. increased variety of proteins.
    b. increased oxygen in the atmosphere.
    c. evolution of DNA.
    d. increased carbon dioxide in the atmosphere.

19. Organisms capable of aerobic metabolism evolved:
    a. prior to photosynthetic organisms.
    b. after photosynthetic organisms.
    c. at the same time as photosynthetic organisms.
    d. as soon as Earth cooled.

20. Humans and chimpanzees diverged from a common ancestor about:
    a. 2 billion years ago.
    b. 5 million years ago.
    c. 50,000 years ago.
    d. 10,000 year ago.

21. In comparing hominids and primates:
    a. both hominids and primates have a tail.
    b. hominids have smaller brains than primates.
    c. both are arboreal.
    d. hominids have more complex social behavior than primates.

22. All of the following apply to *Australopithecus afarensis* except:
    a. their diet consisted of vegetables.
    b. they are classified as hominid.
    c. they displayed sexual dimorphism.
    d. they gave rise to *Homo erectus*.

23. Humans qualify for the phylum Chordata because they have:
    a. upright posture.
    b. spoken language.
    c. a nerve cord.
    d. mammary glands.

24. Evidence for chimpanzees as our closest relatives comes from:
    a. analogous structures.
    b. the absence of a tail in chimpanzees.
    c. similar thumb structure.
    d. similarities in human and chimpanzee DNA.

25. Hominids are characterized by:
    a. complex language skills.
    b. opposable thumbs.
    c. upright posture.
    d. mammary glands.

# Key Concept Review Questions

*Each of the Key Concepts listed at the beginning of this chapter has been rewritten as a question below. After successfully completing the study guide exercises and the Chapter Test, you should be able to answer each of these questions. Refer to the Key Concepts list at the beginning of this chapter to check your answers.*

1. What is evolution?
2. What causes microevolution to occur?
3. What characterizes macroevolution?
4. Who proposed the theory of descent with modification?
5. What are four areas that provide evidence for evolution?
6. What are fossils?
7. Why is the fossil record incomplete?
8. What are homologous structures? What are analogous structures?
9. What are vestigial structures?
10. What characteristic of vertebrate embryos supports the theory of evolution?
11. What information is gained by analyzing biochemical molecules?
12. What is biogeography?
13. What effect do geographic barriers have on evolution?
14. What is the process of natural selection?
15. What two things contribute most to the process of evolution?
16. How can genetic drift and gene flow change populations?
17. How did life on earth probably first begin?
18. What effect did photosynthetic organisms have on the atmosphere of the primitive earth?

19. What is the genus and species for humans?

20. What two characteristics were required in ancestral organisms for the development of humans?

---

# Answer Key

## Sections 22.1, 22.2

**Crossword Puzzle: 1.** selection; **2.** microevolution; **3.** homologous; **4.** extinction; **5.** drift; **6.** radiation; **7.** analogous; **8.** across: flow, down: fossil; **9.** population; **10.** descent over time, genetic modification, unpredictable, natural; **11.** descent with modification; **12.** one; **13.** sediment, volcanic ash; **14.** Macroevolution; **15.** stratification; **16.** lowest; **17.** potassium; **18.** vestigial structures; **19.** biochemical molecules; **20.** diverge; **21.** Biogeography **22.** Pangaea; **23.** mass extinction; **24.** bottleneck effect; **25.** a. homologous, b. descent from a common ancestor, c. divergence allowed each species to evolve along a different path; **26.** a. analogous, b. no, c. they evolved to carry out a similar function

## Sections 22.3, 22.4, 22.5

**1.** b; **2.** i; **3.** f; **4.** k; **5.** h; **6.** a; **7.** d; **8.** j; **9.** c; **10.** e; **11.** g

## Section 22.6

**1.** Animalia; **2.** eukaryotic, multicellular, complex anatomy; **3.** Chordata; **4.** nerve cord, backbone; **5.** Mammalia; **6.** mammary glands; **7.** Primata; **8.** five digits, flat fingernails; **9.** Hominidae; **10.** upright posture, enlarged brain; **11.** *Homo;* **12.** human form; **13.** *Sapiens;* **14.** complex social structure, complex spoken language; **15.** a; **16.** d; **17.** c; **18.** c; **19.** b; **20.** a; **21.** b; **22.** d; **23.** a; **24.** b; **25.** d; **26.** c; **27.** d; **28.** c; **29.** a; **30.** b

## Chapter Test

**1.** a; **2.** c; **3.** e; **4.** c; **5.** b; **6.** b; **7.** a; **8.** c; **9.** c; **10.** c; **11.** c; **12.** a; **13.** b; **14.** b; **15.** c; **16.** b; **17.** b; **18.** b; **19.** b; **20.** b; **21.** d **22.** d; **23.** c; **24.** d; **25.** c

# 23

# Ecosystems and Populations

## Chapter Summary and Key Concepts

*After reading and studying this chapter you should know the following:*

### Sections 23.1, 23.2, 23.3

1. Ecology is the study of the relationships between organisms and their physical environment.

2. Organisms and environments interact as populations, communities, ecosystems, and the biosphere.

3. Characteristics of populations include habitat, range, size, actual growth rate, and potential for growth.

4. The biotic potential of a population is the maximum rate of growth, and is limited by environmental resistance.

5. The population that an ecosystem can support indefinitely is called the carrying capacity.

6. Rapid growth in the human population began in the 1700s when new developments increased the carrying capacity.

7. The fertility rate is the number of children born to each woman; the replacement fertility rate is the average fertility rate required to reach zero population growth.

8. The fertility rate is not the same in all nations; the least developed countries with the fewest resources have the highest fertility rates.

9. An organism's niche is its role in the community. Competition may result when niches for different species overlap.

10. Succession is the natural sequence of change in a community, and is determined by population growth rates, niches occupied by various species, and competition.

11. Succession ends with the establishment of a climax community.

12. An ecosystem consists of living components called the biomass, and non-living components made up of chemical elements and energy.

Chapter 23 *Ecosystems and Populations* 341

**Sections 23.4, 23.5**

13. Energy flow through an ecosystem is governed by the Laws of Thermodynamics.
14. The First Law of Thermodynamics states that energy cannot be created or destroyed.
15. The Second Law of Thermodynamics states that whenever energy changes form or is transferred, some of the energy will be transformed into nonuseful forms.
16. Producers are organisms capable of photosynthesis.
17. Consumers are organisms that must consume food for energy.
18. In an ecosystem, energy flows from the sun to producers, and then to consumers.
19. In a biogeochemical cycle, chemicals cycle between the biomass, exchange pool, and reservoir pool.
20. Important biogeochemical cycles include the cycles of water, carbon, nitrogen, and phosphorus.

# Exercises

*Complete the exercises for each section after you have read and studied the section. If you cannot answer some questions, or answer them incorrectly, return to the chapter and review this information. You may find it helpful to work on only one section at a time. When you have completed all sections, take the Chapter Test as an indicator of your mastery of this topic.*

**23.1  Ecosystems: Communities interact with their environment**

**23.2  Populations: The dynamics of one species in an ecosystem**

**23.3  Human population growth**

## Matching

_____ 1. **ecology**          a. the average fertility rate required to achieve zero population growth

_____ 2. **ecosystem**        b. the maximum rate of growth for a population

_____ 3. **population**       c. nations that are just beginning to industrialize

_____ 4. **community**        d. all the ecosystems of the earth

_____ 5. **biosphere**        e. the number of births per year *minus* the number of deaths per year *divided* by the total population

_____ 6. **habitat**          f. a group of individuals of the same species that occupy the same geographic area

_____ 7. **geographic range**      g. a change in a human population over time

_____ 8. **biotic potential**      h. the study of the relationships between organisms and their physical environment

_____ 9. **carrying capacity**     i. nations with established industry-based economies

_____ 10. **environmental resistance**   j. the populations of all species that occupy the same geographic area

_____ 11. **growth rate**          k. the point at which birth rate equals death rate

_____ 12. **zero population growth**   l. the area over which a species may be found

_____ 13. **fertility rate**       m. a community of organisms and the physical environment in which they live

_____ 14. **replacement fertility rate**   n. the type of location where a species chooses to live

_____ 15. **demographic transition**    o. the number of children born to each woman during her lifetime

_____ 16. **more industrialized countries (MICs)**   p. the population that an ecosystem can support indefinitely

_____ 17. **less industrialized countries (LICs)**   q. environmental factors that limit a species' ability to realize its biotic potential

## Fill-in-the-Blank

*Referenced sections are in parentheses.*

18. A species' _____ is determined by its tolerance for environmental conditions. (23.2)

19. Three things that limit an organism's range are _____ _____ _____, _____ _____, and _____ _____. (23.2)

20. A population stabilizes when it reaches a balance between _____ _____ and _____ _____. (23.2)

21. Factors that kill organisms or prevent them from reproducing contribute to _____ _____. (23.2)

22. Every growing population will eventually reach a point where environmental resistance begins to _____. (23.2)

23. The global human population first began to rise slowly with agricultural development and conditions led to a(n) _____ in environmental resistance and a(n) _____ in carrying capacity. (23.3)

24. In order to reach zero population growth, the human population will have to either _____ the birth rate or _____ death rate. (23.3)

25. In 2000, the human fertility rate was _____. (23.3)

26. The fertility rate is lowest in _____ industrialized countries. (23.3)

27. _____ is the study of human populations. (23.3)

28. The demographic transition of a region is related to its _____ development and the _____ well-being of its citizens. (23.3)

29. In the first stage of demographic transition, a society will have a(n) _____ death rate, and a(n) _____ birth rate. (23.3)

30. In the final stages of demographic transition, the birth rate _____ the death rate. (23.3)

## Short Answer

31. What types of interactions will occur between animals and plants that occupy the same or overlapping habitats?

32. What four factors influence the biotic potential of a species?

33. What contributed to the stability of the human population prior to 5,000 years ago?

34. What historical event coincides with the beginning of rapid growth in the human population?

35. Why is the replacement fertility rate 2.1 rather than 2.0?

36. Which type of countries have the highest fertility rate? What challenges may result from this?

23.4 Communities: Different species living together

23.5 Energy flows through ecosystems

23.6 Chemical cycles recycle molecules in ecosystems

## Matching

____ 1. **niche**     a. a stable community achieved by succession

____ 2. **succession**     b. indicates total amount of energy at each level of an ecosystem or total biomass stored at each level

____ 3. **climax community**     c. a step in the biogeochemical cycle of nitrogen in which bacteria convert nitrates into atmospheric nitrogen gas

____ 4. **biomass**     d. the living components of an ecosystem

____ 5. **producers**     e. the process of converting atmospheric nitrogen gas to ammonia

____ 6. **consumers**     f. carbon-containing compounds obtained from fossils and used as an energy source

____ 7. **photosynthesis**     g. a natural sequence of change in the organisms that dominate a community

____ 8. **herbivores**     h. organisms that cannot synthesize their own food

____ 9. **carnivores**     i. the interaction of multiple food chains

____ 10. **omnivores**     j. consumers that utilize the energy stored in green plants

____ 11. **decomposers**     k. an organism's role in the community

____ 12. **food web**     l. consumers that obtain energy from dead organisms

____ 13. **ecological pyramid**     m. consumers that can derive energy from plants or animals

____ 14. **biogeochemical cycle**     n. the recycling of chemicals that compose living organisms between the organisms and the earth

____ 15. **fossil fuels**     o. the formation of nitrate

____ 16. **nitrogen fixation**     p. organisms capable of photosynthesis

____ 17. **nitrification**     q. consumers that feed on other animals

____ 18. **denitrification**     r. harvesting solar energy to produce organic molecules

## Short Answer

19. Climax communities are very efficient in their utilization of what types of resources?

20. Why are climax communities slow to recover if they are disrupted?

## Labeling

*On the left side of Figure 23.1, label the role each organism plays in the food chain. On the right side of the figure, label the characteristics applicable to each organism. Use the lists provided; each characteristic may be used more than once.*

**Roles**

a. secondary consumer
b. producer
c. tertiary consumer
d. primary consumer

**Characteristics**

a. referred to as an autotroph
b. release $CO_2$ into the atmosphere
c. herbivore
d. performs photosynthesis
e. release $O_2$ into the atmosphere
f. carnivore
g. referred to as a heterotroph
h. least efficient use of energy

_____ 21.

_____ 22.

_____ 23.

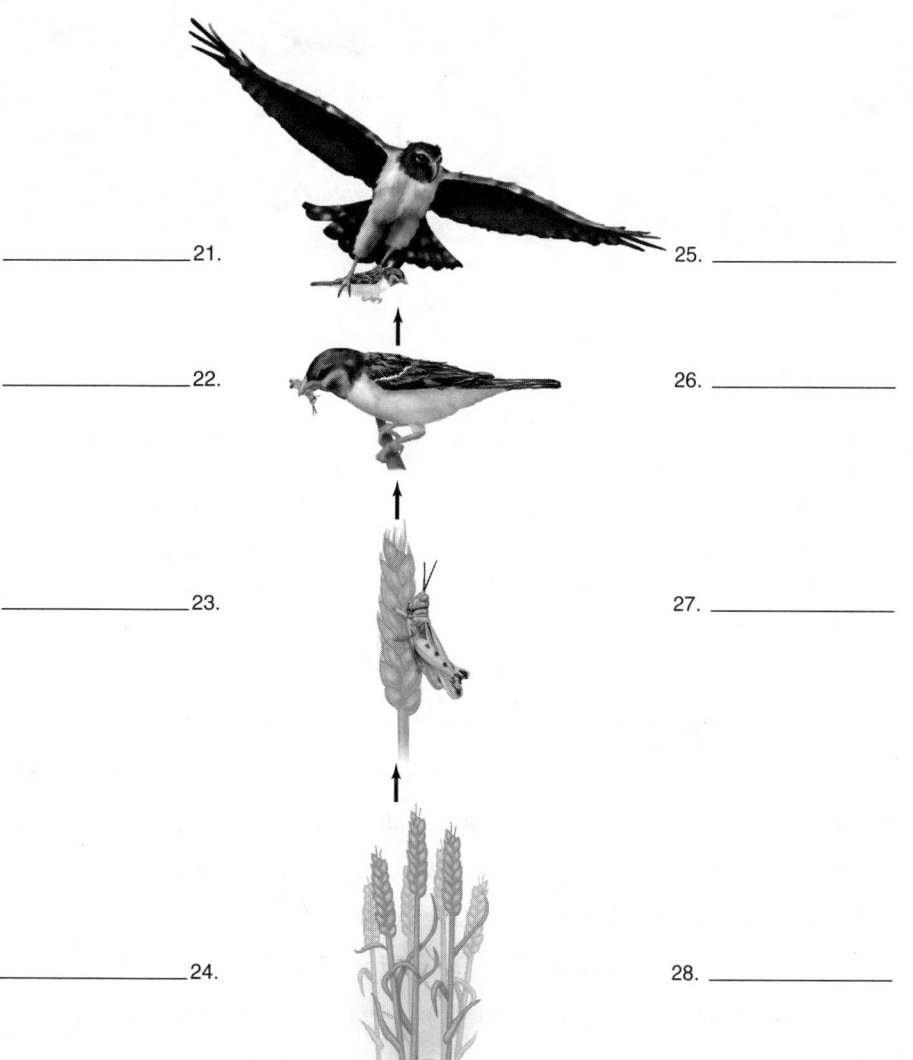

_____ 24.

25. _____

26. _____

27. _____

28. _____

**Figure 23.1**

*Use the phrases below to label Figure 23.2 of the water cycle.*

a. precipitation over land
b. precipitation over ocean
c. ocean to land
d. evaporation from land
e. evaporation from ocean
f. land to ocean

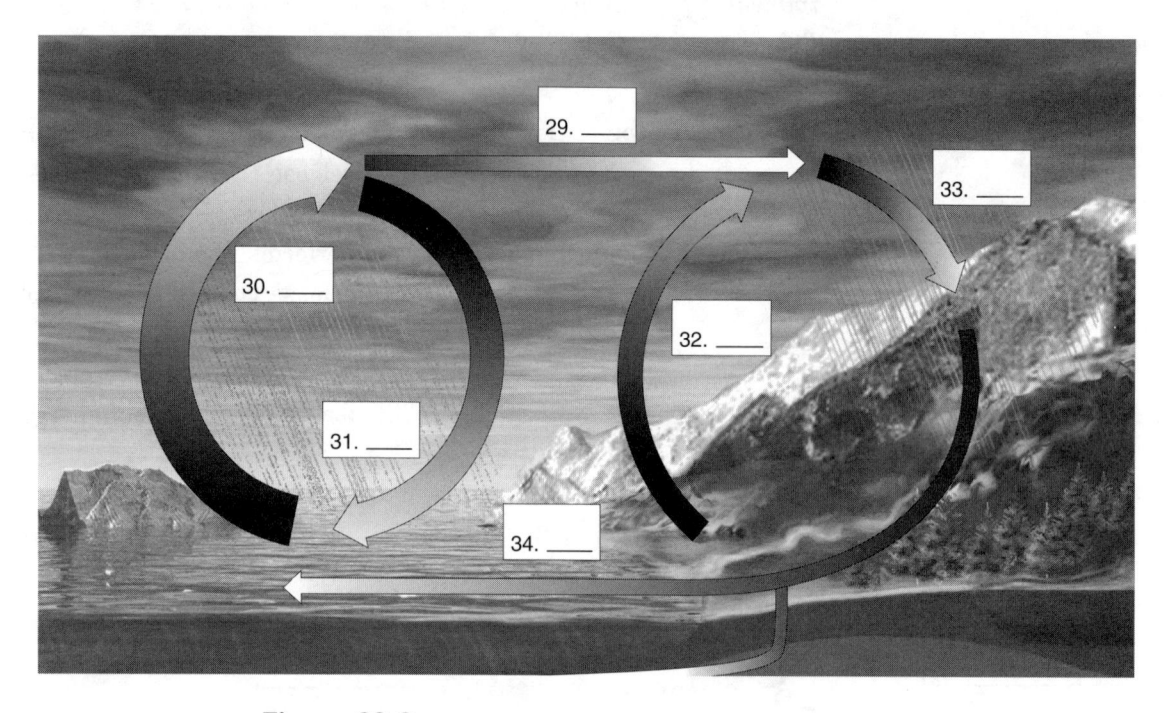

29. _____
30. _____
31. _____
32. _____
33. _____
34. _____

**Figure 23.2**

## Completion

*Use the terms below to complete the paragraph, then repeat this exercise with the terms covered.*

| | | |
|---|---|---|
| carbohydrates | atmosphere | atmospheric $CO_2$ gas |
| organic | bones | aerobic respiration |
| shells | $CO_2$ | solar energy |
| carbohydrates | photosynthesis | aerobic respiration |

Carbon forms the backbone of many (35) _____ molecules and the crystallized structure of (36) _____ and (37) _____. Carbon in living organisms is exchanged with (38) _____. The carbon cycle is dependent on (39) _____ by plants and (40) _____ by plants and animals. During photosynthesis, plants use $CO_2$, (41) _____, and water to form (42) _____, releasing oxygen as a by-product. Plants and animals then utilize (43) _____ with oxygen to break down (44) _____ in plants for energy. This process produces (45) _____, which is released into the (46) _____.

## Fill-in-the-Blank

47. The First Law of Thermodynamics states that energy cannot be created or _____. (23.5)

48. Small organelles called _____ are found in plant cells and house the reactions of photosynthesis. (23.5)

49. Plants contain a light-absorbing pigment called _____. (23.5)

50. In the first stage of photosynthesis, sunlight is absorbed, _____ and _____ are produced, and _____ is released. (23.5)

51. The _____ cycle occurs in the second stage of photosynthesis. (23.5)

52. The main product of photosynthesis is high energy sugar molecules built with _____ from the sun, and carbon atoms from _____ _____. (23.5)

53. During the Calvin cycle of photosynthesis, plants use $CO_2$ from the atmosphere to manufacture _____. (23.5)

54. In a biogeochemical cycle, producers obtain nutrients from the _____ pool, and incorporate them into the _____. (23.6)

55. Nitrogen fixation is carried out by _____. (23.6)

56. All organisms, other than plants, must rely on _____ for usable nitrogen. (23.6)

57. In denitrification, denitrifying bacteria convert some nitrates to _____ _____. (23.6)

58. The phosphorus cycle is called a _____ cycle because phosphorus never enters the atmosphere. (23.6)

## Short Answer

59. Why is energy described as flowing through the ecosystem in one direction, rather than being recycled?

60. Why are nutrients described as cycling through the ecosystem, rather than flowing in one direction?

# Chapter Test

## Multiple Choice

1. Ecology is best defined as the study of:
   a. the cycling of elements in the environment.
   b. the relationship between organisms and their physical environment.
   c. ecological pyramids.
   d. fossil fuels.

2. Which of the following does not limit an organism's geographic range?
   a. competition for resources
   b. reproductive behaviors
   c. extreme or intolerable conditions
   d. physical obstacles and geographic barriers

3. The biotic potential of an organism will usually:
   a. increase without limits.
   b. increase independently of carrying capacity.
   c. not be affected by environmental resistance.
   d. follow an exponential growth curve.

4. One determinant of the biotic potential of a population is:
   a. time required for offspring to reach reproductive maturity.
   b. genetic compatibility.
   c. gene flow and genetic drift.
   d. environmental conditions and geographic barriers.

5. The rapid rise in human population can be attributed to an increase in all of the following except:
   a. industrialization.
   b. availability of antibiotics and vaccines.
   c. environmental resistance.
   d. geographic range.

6. Industrialization raises the —————————— of a human population.
   a. biotic potential
   b. environmental resistance
   c. reproductive rate
   d. carrying capacity

7. Environmental resistance is best defined as:
   a. the ability of an organism to adapt to a wide range of environments.
   b. the ability of an organism to overcome environmental obstacles to reproduction.
   c. factors that limit a species' ability to realize its biotic potential.
   d. the number of populations competing for the same resources.

8. The growth rate of a population is calculated as the number of births per year *minus* the number of deaths per year *divided* by the:
   a. fertility rate.
   b. replacement fertility rate.
   c. current growth rate.
   d. population size.

9. A population can fail to have zero population growth even after reaching the replacement fertility rate due to the:
   a. age structure.
   b. infant mortality rate.
   c. number of elderly in the population.
   d. environmental conditions.

10. The highest growth rates in human populations are seen in:
    a. MICs.
    b. LICs.
    c. there is no distinguishable pattern
    d. countries with higher economic and educational levels.

11. The population that an environment can support indefinitely is called the:
    a. climax community.
    b. carrying capacity.
    c. maximum population.
    d. environmental potential.

12. The carrying capacity of an ecosystem is achieved when:
    a. a balance occurs between biotic potential and environmental resistance.
    b. the reproductive rate of a population declines.
    c. environmental resistance increases.
    d. competitive exclusion increases.

13. Competitive exclusion could not occur without:
    a. two or more populations competing for the same resources.
    b. two or more organisms of the same species competing for the same resources.
    c. more organisms of reproductive age than the environment is capable of supporting.
    d. a change in the physical environment that limits the availability of resources.

14. In photosynthesis, the products of the Calvin cycle include:
    a. $O_2$ and $CO_2$.
    b. $O_2$ and sugar.
    c. sugar and ATP.
    d. sugar and ADP.

15. Chlorophyll is:
    a. a plant cell that carries out photosynthesis.
    b. a species of algae that functions as a producer in an aquatic ecosystem.
    c. a light-absorbing pigment.
    d. a storage organelle for excess carbohydrates in plant cells.

16. Climax communities are:
    a. the most efficient in energy and nutrient utilization.
    b. occupied by one species.
    c. not involved in competition.
    d. only found in certain locations.

17. A constant supply of energy is essential for continued life because the transfer of energy from one organism to another organism is an energy-_____ process.
    a. absorbing
    b. producing
    c. releasing
    d. balanced

18. The chemical matter of the earth is _____ the earth and the biomass.
    a. transferred to
    b. divided between
    c. produced by
    d. recycled between

19. The Second Law of Thermodynamics states that:
    a. energy cannot be created or destroyed.
    b. energy can be changed from one form to another.
    c. energy can be used, or stored in a usable form.
    d. energy transfer is inefficient.

20. Humans can function as all of the following except:
    a. secondary consumers.
    b. tertiary consumers.
    c. omnivores.
    d. autotrophs.

21. The population of consumers at any level of a food chain depends most directly on:
    a. the amount of energy available from the sun.
    b. the biogeochemical cycles of essential chemicals.
    c. the population of consumers directly below it.
    d. the vulnerability of the producers.

22. Molecules and elements in living things cycle between the biomass, a reservoir, and a(n):
    a. producer.
    b. consumer.
    c. exchange pool.
    d. recycling pool.

23. Which of the following would not affect the carbon cycle?
    a. photosynthesis by plants
    b. aerobic respiration by plants and animals
    c. sunlight
    d. atmospheric phosphorus

24. Living things depend on nitrogen fixation:
    a. to detoxify atmospheric nitrogen.
    b. to convert atmospheric nitrogen to a usable form.
    c. to make nitrogen available for aerobic respiration.
    d. to return nitrogen to the atmosphere.

25. The phosphorus cycle is a sedimentary cycle because:
    a. phosphorus cycles through only two pools.
    b. phosphorus is obtained from sediments.
    c. the phosphorus cycle relies on water availability.
    d. phosphorus never enters the atmosphere.

# Key Concept Review Questions

*Each of the Key Concepts listed at the beginning of the study guide has been rewritten as a question below. After successfully completing the study guide exercises and the Chapter Test, you should be able to answer each of these questions. Refer to the Key Concept List at the beginning of this chapter to check your answers.*

1. What is ecology?
2. List four levels of interaction involving organisms with each other or with the environment.
3. List five characteristics of populations.
4. What is the biotic potential of a population and what limits this potential?
5. What is carrying capacity?
6. When did rapid growth in the human population begin, and what made this growth possible?
7. Define fertility rate and replacement fertility rate.
8. In what types of countries are fertility rates the highest?
9. What is an organism's niche, and what happens when the niches for two different species overlap?
10. What is succession, and what factors influence it?
11. What is the culmination of succession?
12. What are the living and non-living components of an ecosystem?
13. What laws govern the flow of energy through an ecosystem?
14. What is the First Law of Thermodynamics?
15. What is the Second Law of Thermodynamics?
16. What are producers?
17. What are consumers?
18. What is the sequence of energy flow in an ecosystem between producers, consumers, and the sun?
19. What happens in a biogeochemical cycle?
20. List four common and important chemicals involved in biogeochemical cycles.

# Answer Key

## Sections 23.1, 23.2, 23.3

**1.** h; **2.** m; **3.** f; **4.** j; **5.** d; **6.** n; **7.** l; **8.** b; **9.** p; **10.** q; **11.** e; **12.** k; **13.** o; **14.** a; **15.** g; **16.** i; **17.** c; **18.** habitat **19.** competition for resources, intolerable conditions, physical obstacles; **20.** biotic potential, carrying capacity; **21.** environmental resistance; **22.** increase; **23.** decrease, increase; **24.** decrease, increase; **25.** 2.65; **26.** more; **27.** Demography; **28.** industrial, economic; **29.** declining, higher; **30.** equals; **31.** They will be in competition for food, shelter, and resources; **32.** the number of offspring produced by each reproducing member, the length of time required for offspring to reach sexual maturity, the ratio of males to females, the number of reproductive members of the population;

**33.** famine, disease, and harsh environmental conditions prevented an increase in the carrying capacity; **34.** the Industrial Revolution; **35.** Not all children will survive to reproductive age; **36.** the LICs; these countries have the fewest resources and the poorest ability to care for a large population

## Sections 23.4, 23.5, 23.6

**1.** k; **2.** g; **3.** a; **4.** d; **5.** p; **6.** h; **7.** r; **8.** j; **9.** q; **10.** m; **11.** l; **12.** i; **13.** b; **14.** n; **15.** f; **16.** e; **17.** o; **18.** c; **19.** energy and nutrients; **20.** Many years are required to establish the complex balance between species and resources that exists in a climax community and it cannot be easily restored; **21.** c; **22.** a; **23.** d; **24.** b; **25.** f,j,k,l; **26.** f,j,k; **27.** g,k; **28.** e,h,i; **29.** c; **30.** e; **31.** b; **32.** d; **33.** a; **34.** f; **35.** organic; **36.** shells; **37.** bones; **38.** atmospheric $CO_2$ gas; **39.** photosynthesis; **40.** aerobic respiration; **41.** solar energy; **42.** carbohydrates; **43.** aerobic respiration; **44.** carbohydrates; **45.** $CO_2$; **46.** atmosphere; **47.** destroyed; **48.** chloroplast; **49.** chlorophyll; **50.** NADPH, ATP, oxygen; **51.** Calvin; **52.** energy, carbon dioxide; **53.** carbohydrates; **54.** exchange, biomass; **55.** bacteria; **56.** plants; **57.** nitrogen gas; **58.** sedimentary; **59.** because some usable energy is lost each time energy is transferred from one organism to another; **60.** because the chemicals that compose living organisms are constantly cycled between organisms and the earth without loss of usable matter

## Chapter Test

**1.** b; **2.** b; **3.** d; **4.** a; **5.** c; **6.** d; **7.** c; **8.** d; **9.** a; **10.** b; **11.** b; **12.** a; **13.** a; **14.** d; **15.** c; **16.** a; **17.** a; **18.** d; **19.** d; **20.** d **21.** c; **22.** c; **23.** d; **24.** b; **25.** d

# 24

# Human Impacts, Biodiversity, and Environmental Issues

## Chapter Summary and Key Concepts

*After reading and studying this chapter you should know the following:*

**Sections 24.1, 24.2, 24.3**

1. Pollution affects our air, water, and land resources.
2. The effects of air pollution include global warming, destruction of the ozone layer, acid rain, and smog production.
3. The primary greenhouse gases are carbon dioxide, methane, and the air pollutants CFCs.
4. The greenhouse effect allows sunlight to pass while trapping heat. This is a normal process and helps to maintain the temperature of the earth.
5. Global warming is an increase in average global temperatures and occurs when excess greenhouse gases are produced.
6. Two human activities that contribute to the potential for global warming are deforestation and the burning of fossil fuels.
7. Ozone occurs near the earth's surface as a result of air pollution, and in the stratosphere where it is protective. CFCs deplete the stratospheric ozone layer.
8. Acid rain is produced when air pollutants combine with water in the atmosphere.
9. Smog occurs when air pollutants react with each other, and usually appears after an area has experienced rapid industrial growth.
10. The effects of human activities on water include unequal use and increasing water pollution.
11. Toxic pollutants in water impact biological magnification, concentrating toxic substances in tertiary consumers.
12. Human activities can lead to pollution of surface water, ground water, and ocean water.
13. Human activities contribute to about 50% of oceanic oil pollution.

14. The expansion of human populations and the search for resources has polluted existing land and led to a reduction in productive land.

15. Challenges that arise when cities replace farmlands include an increased demand on resources and the loss of farming productivity.

16. Land damage in rural areas occurs through stripping of resources by local populations.

**Sections 24.4, 24.5, 24.6**

17. Because energy use is determined by consumption, we have the power to make choices in where we obtain energy and how we use it.

18. Humans have high energy demands and inadequate energy resources.

19. Current energy sources are primarily fossil fuels; additional energy sources may include nuclear energy, hydroelectric power, biomass fuel, and solar energy.

20. A sustainable world is one in which humans and the ecosystems are maintained.

# Exercises

*Complete the exercises for each section after you have read and studied the section. If you cannot answer some questions, or answer them incorrectly, return to the chapter and review this information. You may find it helpful to work on only one section at a time. When you have completed all sections, take the Chapter Test as an indicator of your mastery of this topic.*

**24.1  Pollutants impair air quality**

**24.2  Pollution jeopardizes scarce water supplies**

**24.3  Pollution and overuse damage the land**

## Matching

\_\_\_\_  1. **pollutants**  a. a condition that occurs when a warm upper layer of air traps a cooler air mass containing smog beneath it

\_\_\_\_  2. **greenhouse effect**  b. the accumulation of organic and inorganic nutrients leads to the growth of plant life and death of animal life

\_\_\_\_  3. **global warming**  c. the transformation of marginal lands into near-desert conditions

\_\_\_\_  4. **deforestation**  d. deep underground reservoirs of fresh water

\_\_\_\_  5. **ozone**  e. chemicals in the environment that have adverse effects on living organisms

\_\_\_\_  6. **acid rain**  f. a condition where toxic pollutants become more concentrated in the tissue of animals higher in the food chain

_____ 7. **smog**  g. when sunlight is allowed to pass through the atmosphere but most of the heat is trapped

_____ 8. **thermal inversion**  h. removing trees from large areas of land

_____ 9. **aquifers**  i. a raise in the average global temperature

_____ 10. **eutrophication**  j. $O_3$ found in two places in the atmosphere

_____ 11. **biological magnification**  k. a brown or gray layer formed in the atmosphere when air pollutants react with each other

_____ 12. **desertification**  l. produced when air pollutants combine with water in the atmosphere

## Short Answer

*Refer to Figure 24.1 to answer the following questions.*

13. Which pathways of the carbon cycle have been affected by human activity in such a way that atmospheric levels of $CO_2$ are increasing?

14. Describe the activity associated with each pathway and how it is causing $CO_2$ levels to rise.

15. Does this increase in $CO_2$ levels increase or decrease the greenhouse effect?

16. What effect can this rising $CO_2$ level have on the temperature of the earth?

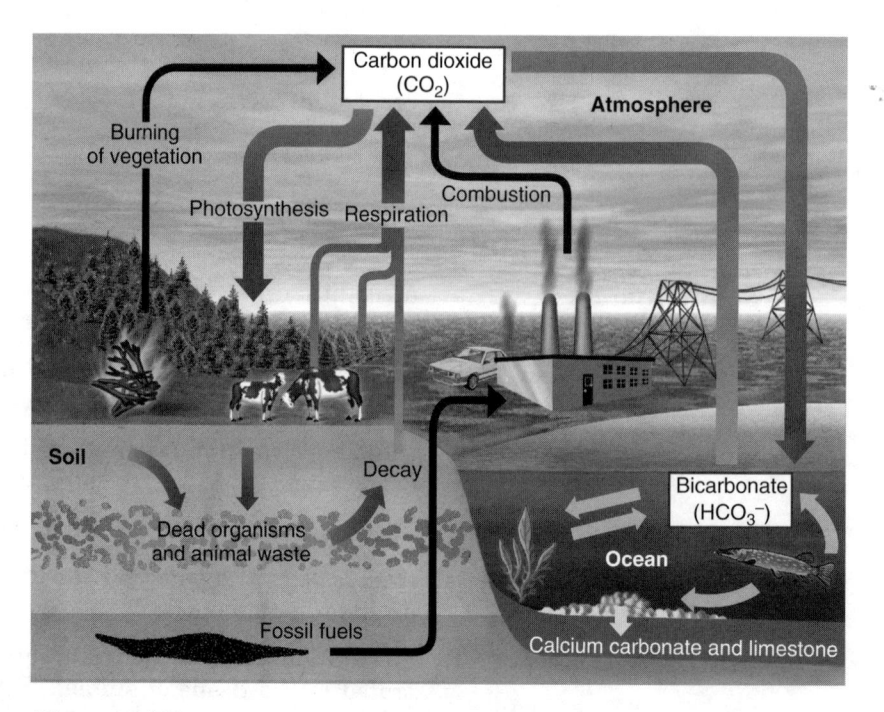

**Figure 24.1**

## Word Choice

*Circle the answer that correctly completes the sentence.*

17. Ozone in the __(troposphere/stratosphere)__ helps to protect the earth from UV radiation.

18. Ozone in the __(troposphere/stratosphere)__ is a pollutant formed when oxygen reacts with automobile exhaust and industrial pollution.

19. CFCs __(increase/decrease)__ the level of ozone in the stratosphere.

20. CFCs __(increase/decrease)__ global warming.

21. Acid rain may form when __(chlorine/sulfur dioxide)__ is released into the air.

22. Nitrogen oxides and hydrocarbons are the primary contributors to __(global warming/smog formation)__.

23. Freshwater comprises __(less than/more than)__ 1% of the earth's total amount of water.

24. Water pollutants that are organic nutrients __(increase/decrease)__ the growth rate of bacteria, resulting in a(n) __(increased/decreased)__ availability of oxygen for other aquatic organisms.

25. Eutrophication __(converts freshwater into/protects freshwater from becoming)__ a marsh.

26. Most toxic pollutants __(can/cannot)__ be degraded by biological decomposition.

27. In biological magnification, toxic substances become __(more/less)__ concentrated in the tissues of animals higher up in the food chain.

28. Many of the pollutants that affect surface water __(do/do not)__ pollute groundwater.

29. Desertification __(increases/decreases)__ the amount of available productive land.

## Fill-in-the-Blank

*Referenced sections are in parentheses.*

30. Air pollution impacts the environment in four areas:
    _____ _____,
    _____ _____,
    _____ _____, and
    _____ _____. (24.1)

31. CFCs produce _____ atoms that convert ozone into _____. (24.1)

32. Sewage treatment plants, food-packing plants, and paper mills produce _____ _____ that pollute water. (24.2)

33. The pollution of water by inorganic nutrients causes an increase in the growth of _____. (24.2)

34. Heat pollution affects water by decreasing the amount of _____ that water can carry. (24.2)

35. One source of heat pollution is _____ _____. (24.2)

36. Pollution of groundwater is a concern for human health because it is often a source of _____ _____. (24.2)

37. The EPA estimates that up to _____ percent of all water systems and rural wells are polluted. (24.2)

38. About _____ percent of oceanic oil pollution is caused by man. (24.2)

24.4 Energy: many options, many choices

24.5 Human impacts are creating a biodiversity crisis

24.6 Toward a sustainable future

## Completion

*Fill in the table below, listing the disadvantages of utilizing various sources of energy.*

| Energy Source | Disadvantages |
|---|---|
| Fossil Fuels | 1. |
|  | 2. |
|  | 3. |
|  | 4. |
| Nuclear Energy | 5. |
|  | 6. |
|  | 7. |
| Hydroelectric Power | 8. |
| Biomass Fuels | 9. |
|  | 10. |

## Short Answer

11. Solar power is a renewable energy source. What currently prevents solar power from becoming our primary energy source? (24.4)

12. What human activities have contributed to a reduction in biodiversity? (24.5)

13. What are four things we can do to work toward creating a sustainable world? (24.5)

# Chapter Test

## Multiple Choice

1. Which of the following is not a primary component of the air we breathe?
   a. nitrogen
   b. oxygen
   c. carbon dioxide
   d. magnesium

2. The greenhouse effect:
   a. prevents sunlight from passing.
   b. allows heat to escape.
   c. is important in maintaining normal surface temperatures on the earth.
   d. is not involved in global warming.

3. The two main human activities that raise the level of $CO_2$ in the atmosphere are:
   a. deforestation.
   b. desertification.
   c. burning of fossil fuels.
   d. a and c.
   e. b and c.

4. Ozone in the atmosphere is:
   a. beneficial in the troposphere and harmful in the stratosphere.
   b. beneficial in the stratosphere and harmful in the troposphere.
   c. beneficial in both the troposphere and the stratosphere.
   d. harmful in both the troposphere and the stratosphere.

5. CFCs contribute to global warming by:
   a. releasing chlorine atoms that produce more ozone.
   b. destroying the chlorine atoms that normally produce ozone.
   c. releasing chlorine atoms that destroy ozone and produce oxygen.
   d. releasing oxygen atoms that convert ozone into chlorine atoms.

6. Which of the following does not contribute directly to the formation of acid rain?
   a. burning high-sulfur coal
   b. automobile exhaust
   c. burning oil
   d. deforestation

7. Acid rain is produced when:
   a. holes in the ozone layer affect atmospheric gases.
   b. sulfur dioxide and nitrogen oxides combine with water vapor in the air.
   c. water in the atmosphere ionizes.
   d. industrial plants release acid vapors into the air.

8. The two factors that contribute most to smog are:
   a. acid rain and the burning of fossil fuels.
   b. the burning of fossil fuels and automobile exhaust.
   c. an increase in greenhouse gases and automobile exhaust.
   d. eutrophication and automobile exhaust.

9. What condition can be worsened by a thermal inversion?
   a. ozone depletion
   b. smog
   c. acid rain
   d. global warming

10. Fresh water comprises _____ of the earth's total water.
    a. less than 1%
    b. less than 5%
    c. more than 10%
    d. more than 50%

11. The availability of fresh water in the world:
    a. is used primarily by populations of more industrialized countries.
    b. is used primarily by populations of less industrialized countries.
    c. is equally distributed throughout most nations.
    d. has little or no effect on human carrying capacity.

12. Water pollutants that are inorganic nutrients:
    a. cause prolific growth of algae.
    b. cause prolific growth of bacteria.
    c. are degraded by bacteria.
    d. arise from sewage treatment plants.

13. Eutrophication:
    a. refers to the rapid growth of animal life in a shallow body of water.
    b. results from a depletion of organic nutrients.
    c. is accelerated by water pollution.
    d. converts dry land into marsh.

14. In biological magnification, _____ become more concentrated in animals that are _____.
    a. toxic substances, higher up in the food chain
    b. toxic substances, lower in the food chain
    c. nutrients, higher up in the food chain
    d. nutrients, lower in the food chain

15. Which of the following does not contribute directly to water pollution?
    a. disease-causing organisms
    b. heat
    c. sediment
    d. CFCs

16. Heat pollution in water:
    a. increases the amount of oxygen in the water.
    b. decreases the oxygen demand of some aquatic organisms.
    c. may lead to suffocation of some aquatic organisms.
    d. protects freshwater from organic contaminants.

17. It is difficult to determine the full effects of groundwater contamination because:
    a. the effects may be confused with other causes of human disease.
    b. it is difficult to detect the presence of radioactive waste.
    c. groundwater is protected from most contaminants by its location.
    d. many effects of groundwater contamination are never reported.

18. About half of the oil that enters the ocean every year comes from:
    a. oil disposal on land that washes out to sea.
    b. accidents at sea.
    c. natural seepage.
    d. acid rain.

19. Which of the following contributes least to the damage and overuse of land?
    a. reliance on the land as a resource in rural communities
    b. expansion of cities
    c. urban poverty
    d. war

20. When land becomes unusable for agricultural purposes as a result of approaching desert-like conditions, this is called:
    a. desertification.
    b. erosion.
    c. heat pollution.
    d. eutrophication.

21. Which of the following is a true statement about fossil fuels?
    a. They are one fuel source that does not have an environmental cost.
    b. Use of fossil fuels may reduce the level of greenhouse gases.
    c. Use of fossil fuels may contribute to acid rain.
    d. They are a nonrenewable energy source.

22. Examples of biomass fuels include:
    a. wood.
    b. coal.
    c. oil.
    d. water.

23. The original renewable energy source for our world is:
    a. hydroelectric power.
    b. solar energy.
    c. human kinetic energy.
    d. biomass fuels.

24. Human activities that impact biodiversity include all of the following except:
    a. heat pollution.
    b. depletion of natural resources.
    c. farming.
    d. deforestation.
    e. all forms of pollution and resource depletion impact biodiversity

25. Which of the following would not contribute to the establishment of a sustainable world?
    a. recycling
    b. reducing consumption
    c. increasing the fertility rate
    d. reducing poverty

# Key Concept Review Questions

*Each of the Key Concepts listed at the beginning of the study guide have been rewritten as a question below. After successfully completing the study guide exercises and the Chapter Test, you should be able to answer each of these questions. Refer to the Key Concept List at the beginning of this chapter to check your answers.*

1. What areas of our environment are affected by pollution?
2. What are the four primary effects of air pollution?
3. What are the primary greenhouse gases?

4. What is the greenhouse effect? Is it a normal process? What aspect of the environment do greenhouse gases help to maintain?

5. What is global warming? What may cause global warming?

6. What two human activities contribute to the potential for global warming?

7. Where is ozone found? In which area is ozone beneficial? What chemicals contribute to depletion of the ozone layer?

8. What causes acid rain?

9. What causes smog? When does it usually appear?

10. List two ways in which human activities affect water as a resource.

11. What happens in the process of biological magnification?

12. What sources of water can by polluted by human activities?

13. What percentage of oceanic oil pollution is the result of human activity?

14. What effect have human activities had on existing land and available productive land?

15. List two challenges that arise when cities replace farmlands.

16. What is the primary cause of land damage in rural areas?

17. Why do we have the power to make choices in where we obtain energy and how we use it?

18. How would you characterize the energy demands of human populations?

19. What is currently the primary energy source for human populations? List four additional energy sources.

20. What is a sustainable world?

# Answer Key

## Sections 24.1, 24.2, 24.3

1.e; 2.g; 3.i; 4.h; 5.j; 6.l; 7.k; 8.a; 9.d; 10.b; 11.f; 12.c; 13. burning of vegetation, combustion; 14. burning of vegetation releases $CO_2$ from the vegetation and reduces the number of plants available for $CO_2$ uptake in photosynthesis, combustion releases $CO_2$ from fossil fuels; 15. increase; 16. temperatures may increase; 17. stratosphere; 18. troposphere; 19. decrease; 20. increase; 21. sulfur dioxide; 22. smog formation; 23. less than; 24. increase, decreased; 25. converts freshwater into; 26. cannot; 27. more; 28. do; 29. decreases; 30. global warming, ozone destruction, acid rain, smog production; 31. chlorine, oxygen; 32. organic nutrients; 33. algae; 34. oxygen; 35. power plants; 36. drinking water; 37. 50; 38. 50

## Sections 24.4, 24.5, 24.6

**1.** retrieving and transporting them has an environmental cost; **2.** may contribute to greenhouse gases; **3.** coal contains sulfur and may lead to formation of acid rain; **4.** fossil fuels are a nonrenewable resource; **5.** expensive; **6.** potentially dangerous; **7.** generates hazardous waste; **8.** environmental cost; **9.** contribute to pollution; **10.** not always available; **11.** Solar power technology is not capable of meeting the energy demand; **12.** pollution, destruction of habitats, exploitation of resources, see sec. 24.5 for additional possible answers; **13.** consume less, recycle more, lower the worldwide fertility rate, reduce world poverty

## Chapter Test

**1.**d; **2.**c; **3.**d; **4.**b; **5.**c; **6.**d; **7.**b; **8.**b; **9.**b; **10.**b; **11.**a; **12.**e; **13.**c; **14.**a; **15.**d; **16.**c; **17.**a; **18.**c; **19.**c; **20.**a; **21.**c; **22.**a; **23.**b; **24.**e; **25.**c